Letts

D0512069

GCSE success

ESSENTIALS

MATHEMATICS

AUTHOR
Fiona Mapp

Contents

Introduction

Introduction   4

Quick Reference

Time, calendar, angle facts   6
Comparing metric and imperial units, circle theorems   7
Formulae: foundation tier   8
Formulae: intermediate tier   9
Formulae: higher tier   10
Using a calculator   12

Tricky Topics: Intermediate

Number Percentage change   14
Reverse percentage problems   15
Standard index form   16
Indices   17
Algebra Multiplying out brackets   18
Factorisation   19
Rearranging formulae   20
Finding the equation of a straight line   21
Solving equations   22
Solving inequalities   23
Shape, space and measures Pythagoras' theorem   24
Trigonometry   25
Similarity   26
Loci   27
Handling data Averages of grouped data   28
Moving averages   29
Cumulative frequency graphs   30
Median and interquartile range   31

Tricky Topics: Higher

Number Proportion   32
Surds   33
Indices   34
Upper and lower bounds   35
Algebra Solving quadratic equations   36
Algebraic fractions   38
Solving algebraic fractions   39
Rearranging formulae   40
Intersection of a circle and a line   41
Functions and transformations   42
Shape, space and measures Trigonometry in three dimensions   44
The sine and cosine rules   46

	Arc length	48
	Sector area	49
	Vector geometry	50
Handling data	Tree diagrams	51
	Histograms	52

Non-calculator techniques

Multiply and divide		54
Fractions	Adding and subtracting	56
	Multiplying and dividing	57
Percentages and estimating		58
Standard index form		59

Coursework

A close look at AO1: Using and applying mathematics	60
Assessment objective 1	60
Using and applying mathematics: Assessment criteria	62
Tips for carrying out mathematical investigations	64
For grades A and A*	66
Practical work	67
Using examples in practical work	68
A practical project	70
Recording and tabulating results	71
Using ICT: spreadsheets	72
Continuing the investigation	73
A close look at AO4: Handling data	74
The data-handling cycle	75
Handling data: Assessment criteria	76
Tips for carrying out a handling data project	
Choosing a project	78
Formulating a hypothesis	79
Making a plan	80
Collecting the data	82
Recording the data	83
Presenting the data	84
Analyse and interpret	85
Analyse, interpret and draw conclusions	86
For grades A and A*	87
Using ICT	88
Writing up coursework	89
Your final mark	90

Reference

Glossary	91

This book complements the **GCSE Success Guides** and **GCSE Success Guide Questions and Answers**. It includes useful tips, with emphasis on areas that most students find difficult. It is divided into several sections.

Quick reference

Keep the book at hand during mathematics lessons, while doing your homework and when revising for your GCSE Mathematics examination. It contains formulae that you need to learn and useful facts that you need to memorise.

Tricky topics

Each year the examiners' report, written about the GCSE Mathematics examination, highlights areas in which students did well in the examination. It also details those topics that students found difficult. Several of these topics come up each year. The aim of this section is therefore to highlight those particularly difficult areas. Most of these topics are targeted at Grade B of the Intermediate tier and Grade A/A* of the Higher tier. For more help and practice, use the appropriate **GCSE Success Guide** and **GCSE Success Guide Questions and Answers**.

NON-CALCULATOR TECHNIQUES

It is important that you achieve the best possible marks on both of the examination papers. Some students find the non-calculator paper more difficult as they are less confident when they cannot use a calculator. This section highlights some of the basic mathematics skills that you need. If you sharpen up these skills, not only will you achieve higher marks but it will give you more confidence too.

Coursework

Since the coursework element of GCSE Mathematics counts for 20% of the final mark, it is important to get as high a mark as possible. This section has been written to give you guidance when you are tackling coursework. The first section is on AO1 – the investigation/practical work. As most students achieve relatively high marks in this topic the content includes quick tips. For most students, the *handling data* project, which was examined for the first time in summer 2003, is still a difficult one in which to achieve high marks. This section gives useful hints, tips and examples to help you produce a good Handling data project.

Keys to success

HOW TO REVISE MATHEMATICS

- Here are a few tips to help you with your revision.

  Mathematics should be revised actively – doing, not merely reading.

Planning

- Find out the dates of your first mathematics examination.

- Make a revision timetable. Include the dates and times of the examinations.

- After completing a topic in school, go through it again in the **GCSE Success Guide**. Copy the main points, results or formulae into a notebook or use a highlighter to emphasise them.

- Try to write out these key points from memory. Check what you have written and see if there are any differences.

Revising

- Revise in short bursts of about 30 minutes, followed by a short break.

- Learn facts from your exercise books, notebooks and the **GCSE Success Guide**.

- Memorise any formulae you need to learn. Learning with a friend makes it easier and more fun.

- Look through and make sure that you understand the examples in the **GCSE Success Guide**.

- Do the multiple-choice and quiz-style questions in the **GCSE Success Guide Questions and Answers** to see how much you know and check your solutions.

- Once you feel confident that you know the topic, do the exam-style questions in the **GCSE Success Guide Questions and Answers**. Highlight the key words in each question, plan your answer and then, when you have finished, go back and check that you have actually answered the question.

- Make a note of any topics that you do not understand and go back through the notes again.

LEARN YOUR LINGO

The glossary at the end of the book will help you when you come across unfamiliar terms in mathematics lessons or when doing your homework.

Time

60 seconds = 1 minute
60 minutes = 1 hour
24 hours = 1 day
7 days = 1 week
12 months = 1 year
365 days = 1 year
366 days = 1 leap year
10 years = 1 decade
100 years = 1 century

Calendar

There are 28 days in February (29 in a leap year).

There are 30 days in April, June, September, November.

There are 31 days in January, March, May, July, August, October, December.

Angle facts

Angles on a straight line add up to 180°.

Angles at a point add up to 360°.

Vertically opposite angles are equal.
$a = c$
$b = d$

Alternate angles are equal.

Supplementary angles add up to 180°.
$c + d = 180°$

Angles in a triangle add up to 180°.

Angles in a quadrilateral add up to 360°.

An exterior angle of a triangle equals the sum of the two opposite interior angles.
$d = a + b$

Corresponding angles are equal.

Comparing metric and imperial units

LENGTH
2.5 cm ≈ 1 inch
30 cm ≈ 1 foot
1 m ≈ 39 inches
8 km ≈ 5 miles

Tip
These comparisons are only approximate.

WEIGHT
25 g ≈ 1 ounce
1 kg ≈ 2.2 pounds

CAPACITY
1 litre ≈ $1\frac{3}{4}$ pints
4.5 litres ≈ 1 gallon

Circle theorems

The angle subtended at the centre of a circle is twice the angle subtended at the circumference, by the same arc.

The angles subtended by the same arc in the same segment are equal.

The angles in a semicircle are right angles.

The tangent to a circle is at 90° to the radius at the point of contact.

radius

tangent

The opposite angles in a cyclic quadrilateral add up to 180°.

From any point outside the circle, two tangents can be drawn. The lengths of the tangents, from the exterior point to the point of contact, are equal.

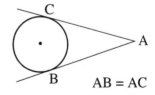

AB = AC

The angle between a tangent and a chord is equal to any angle subtended by that chord in the alternate segment of the circle. This is known as the alternate segment theorem.

a = b

Formulae sheet: Foundation tier

For the Foundation tier, learn these formulae and the multiplication tables.

Area and volume

Area of triangle $= \frac{1}{2} \times$ base $\times$ height

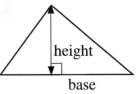

Circumference of circle $= \pi \times$ diameter
$$= 2 \times \pi \times \text{radius}$$

Area of circle $= \pi \times (\text{radius})^2$

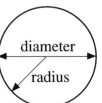

Volume of cuboid
$= $ length $\times$ width $\times$ height

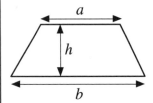

*Area of trapezium
$= \frac{1}{2}(a + b)h$

Tip The formulae marked with asterisks(*) are given on the front of the examination paper.

Multiplication table

	1	2	3	4	5	6	7	8	9	10	11	12
1	1	2	3	4	5	6	7	8	9	10	11	12
2	2	4	6	8	10	12	14	16	18	20	22	24
3	3	6	9	12	15	18	21	24	27	30	33	36
4	4	8	12	16	20	24	28	32	36	40	44	48
5	5	10	15	20	25	30	35	40	45	50	55	60
6	6	12	18	24	30	36	42	48	54	60	66	72
7	7	14	21	28	35	42	49	56	63	70	77	84
8	8	16	24	32	40	48	56	64	72	80	88	96
9	9	18	27	36	45	54	63	72	81	90	99	108
10	10	20	30	40	50	60	70	80	90	100	110	120
11	11	22	33	44	55	66	77	88	99	110	121	132
12	12	24	36	48	60	72	84	96	108	120	132	144

Formulae sheet: Intermediate tier

For the Intermediate tier, you need these formulae.

Area of triangle

$= \frac{1}{2} \times$ base $\times$ height

Circumference of circle

$= \pi \times$ diameter

$= 2 \times \pi \times$ radius

Area of circle $= \pi \times (\text{radius})^2$

Area of parallelogram

$=$ base $\times$ height

*Area of trapezium

$= \frac{1}{2}(a + b)h$

Volume of cuboid

$=$ length $\times$ width $\times$ height

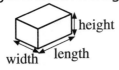

Volume of cylinder $= \pi r^2 h$

*Volume of prism

$=$ (area of cross-section) $\times$ length

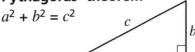

Pythagoras' theorem

$a^2 + b^2 = c^2$

Trigonometry

$\sin \theta = \dfrac{\text{opposite}}{\text{hypotenuse}}$

$\cos \theta = \dfrac{\text{adjacent}}{\text{hypotenuse}}$

$\tan \theta = \dfrac{\text{opposite}}{\text{adjacent}}$

COMMON RELATIONSHIPS

$\text{Speed} = \dfrac{\text{distance}}{\text{time}}$

$\text{Time} = \dfrac{\text{distance}}{\text{speed}}$

Distance $=$ speed $\times$ time

$\text{Density} = \dfrac{\text{mass}}{\text{volume}}$

$\text{Volume} = \dfrac{\text{mass}}{\text{density}}$

Mass $=$ volume $\times$ density

Formulae sheet: Higher tier

For the Higher tier you need these formulae.

Area of triangle
$= \frac{1}{2} \times$ base $\times$ height

Area of parallelogram
$=$ base $\times$ height

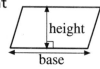

Area of trapezium
$= \frac{1}{2}(a + b)h$

Volume of cuboid
$=$ length $\times$ width $\times$ height

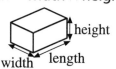

*Volume of prism
$=$ (area of cross-
    section) $\times$ length

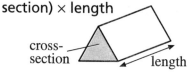

Volume of cylinder
$= \pi r^2 h$

Area of curved surface
of cylinder $= 2\pi rh$

Surface area
of cylinder
$= 2\pi rh + 2\pi r^2$

*Volume of sphere
$= \frac{4}{3}\pi r^3$

*Surface area
of sphere
$= 4\pi r^2$

*Volume of cone
$= \frac{1}{3}\pi r^2 h$

*Area of curved
surface of
cone $= \pi rl$

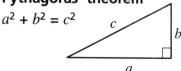

Pythagoras' theorem
$a^2 + b^2 = c^2$

Circumference of circle
$= \pi \times$ diameter
$= 2 \times \pi \times$ radius

Area of circle $= \pi \times$ (radius)^2

Tip The
formulae
marked with
asterisks(*) are
given on the front
of the examination
paper.

TRIGONOMETRY

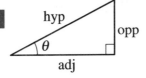

$$\sin \theta = \frac{\text{opposite}}{\text{hypotenuse}} \qquad \cos \theta = \frac{\text{adjacent}}{\text{hypotenuse}} \qquad \tan \theta = \frac{\text{opposite}}{\text{adjacent}}$$

In any triangle ABC:

*Sine rule $\dfrac{a}{\sin A} = \dfrac{b}{\sin B} = \dfrac{c}{\sin C}$

*Cosine rule $a^2 = b^2 + c^2 - 2bc\cos A$

$\cos A = \dfrac{b^2 + c^2 - a^2}{2bc}$

*Area of triangle $= \frac{1}{2}ab \sin C$

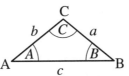

QUADRATIC EQUATIONS

*The solutions of $ax^2 + bx + c = 0$, where $a \neq 0$, are given by

$$x = \frac{-b \pm \sqrt{b^2 - 4ac}}{2a}$$

COMMON RELATIONSHIPS

Speed $= \dfrac{\text{distance}}{\text{time}}$	**Density** $= \dfrac{\text{mass}}{\text{volume}}$
Time $= \dfrac{\text{distance}}{\text{speed}}$	**Volume** $= \dfrac{\text{mass}}{\text{density}}$
Distance = speed × time	**Mass** = volume × density

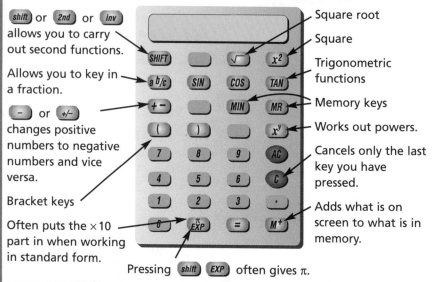

Using a calculator

There are many different makes of calculator. It is important that you know how your own calculator works.

Important calculator keys

This imaginary calculator shows some of the important keys. Make sure you know where these keys are on your own calculator.

Practise using your own calculator. Make sure you know where these keys are.

shift or **2nd** or **inv** allows you to carry out second functions.

Allows you to key in a fraction.

– or **+/–** changes positive numbers to negative numbers and vice versa.

Bracket keys

Often puts the ×10 part in when working in standard form.

Square root

Square

Trigonometric functions

Memory keys

Works out powers.

Cancels only the last key you have pressed.

Adds what is on screen to what is in memory.

Pressing **shift** **EXP** often gives π.

EXAMPLE

Use a calculator to work out $\dfrac{3.7 + 2.5}{4.6 - 3.1}$.

Using brackets, key in:

(3 . 7 + 2 . 5) ÷ (4 . 6 − 3 . 1) =

Check with your own calculator.

Using the memory keys, key in:

4 . 6 − 3 . 1 = MIN 3 . 7 + 2 . 5 = ÷ MR =

The answer is 4.13 (correct to 2 d.p.)

Calculating powers

The y^x or x^y key is used for calculating powers such as 5^6.

- Use the power key on the calculator to work out 5^6.
- Write down the calculator keys you used.
- Check that you obtain the answer 15 625.

Now try writing down the key sequences you would use for these calculations. Check that you get the answers right.

$$\frac{3.2 \times 4.3^2}{(1.2)^3} = 34.24 \qquad\qquad 8^{\frac{2}{3}} \times 4^3 = 256$$

Standard form and the calculator

To key a number in standard form into a calculator, use the $\boxed{EXP}$ key. On some calculators the key is $\boxed{EE}$. Make sure you check your calculations.

EXAMPLE

$6.2 \times 10^5 \times 3.2 \times 10^{-12}$ can be keyed as:

$\boxed{6}\ \boxed{.}\ \boxed{2}\ \boxed{EXP}\ \boxed{5}\ \boxed{\times}\ \boxed{3}\ \boxed{.}\ \boxed{2}\ \boxed{EXP}\ \boxed{1}\ \boxed{2}\ \boxed{+-}$

The answer is 1.984×10^{-6}.

Most calculators do not show standard form correctly on the display.

$\boxed{4.27\ 9}$ means 4.27×10^9.

Remember to put the $\times 10$ part in. A common mistake is to leave it out.

For more help see Higher Success Guide page **17**
Intermediate Success Guide page **25**

Percentage change

For this topic, you might be asked to work out the percentage increase, percentage decrease, percentage profit or loss.

$$\text{Percentage change} = \frac{\text{change}}{\text{original}} \times 100\%$$

Tip The most common error is dividing by the new value and not by the original!

EXAMPLE 1

A house is bought for £98 000.
Two years later it is sold for £147 000.
What is the percentage profit?

Profit = £147 000 − £98 000 = £49 000

Percentage profit
$= \frac{49\,000}{98\,000} \times 100\%$

$= 50\%$

EXAMPLE 2

A car is bought for £14 250. Two years later it is sold for £9670. Work out the percentage loss.

Loss = £14 250 − £9670 = £4580

Percentage loss $= \frac{4580}{14\,250} \times 100\%$

$= 32\%$

For more help see **Intermediate Success Guide** page **12**

Reverse percentage problems

These are tricky. You are usually asked to find the original quantity!

EXAMPLE

The cost of a meal is £112.80 including VAT at 17.5%. Work out the price of the meal before VAT is added.

The price of the meal is 100% + 17.5% = 117.5% of the price without VAT.

$$117.5\% = \frac{117.5}{100} = 1.175$$

$$1.175 \times \text{(original price)} = £112.80$$

$$\text{Original price} = \frac{£112.80}{1.175} = £96$$

Tip Always check that your answer sounds sensible. In this example the answer must be smaller than the starting amount.

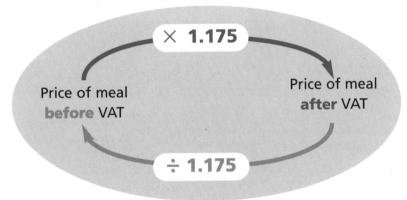

× **1.175**

Price of meal **before** VAT

Price of meal **after** VAT

÷ **1.175**

For more help see Intermediate Success Guide page **13**

Standard index form

You can use standard index form to write very large numbers or very small numbers in a simple way. Any number can be written in standard form as:

$$a \times 10^n$$

where the value of a is between 1 and 10 and n is an integer or whole number.

$$1 \leqslant a < 10$$

The value of n is the number of places that you would have to move the digits to the left to return the number to its original value.

EXAMPLES

$2\,710\,000 = 2.71 \times 10^6$
$32\,000 = 3.2 \times 10^4$
$0.0036 = 3.6 \times 10^{-3}$
$0.000\,005\,7 = 5.7 \times 10^{-6}$

TAKE CARE!

Common mistakes when answering standard form questions include:

- reading a calculator display incorrectly
  e.g. $\boxed{3.6 \quad 8}$ may be written as 3.6^8 instead of 3.6×10^8.
- forgetting to write the answer in standard form, particularly on the non-calculator paper
  e.g. $(3 \times 10^4) \times (4 \times 10^5) = (3 \times 4) \times (10^4 \times 10^5)$
  $$= 12 \times 10^9$$
  $$= 1.2 \times 10^{10}$$

For more help see **Intermediate Success Guide** pages **24** and **25**

Indices

$$a^b$$

base ⟶ a^b ⟵ power or index

This is read as 'a to the power b'. For example, 2^3 means '2 to the power 3' or $2 \times 2 \times 2 = 8$.

The laws of indices apply for numbers and for letters.

$a^n \times a^m = a^{n+m}$

$a^n \div a^m = a^{n-m}$

$(a^n)^m = a^{n \times m}$

$a^0 = 1$

$a^1 = a$

$a^{-1} = \dfrac{1}{a^1}$

Tip Remember that the base has to be the same if the laws of indices are applied.

EXAMPLES

$4^2 \times 4^5 = 4^7$

$5^6 \div 5^2 = 5^4$

$(2^5)^3 = 2^{15}$

$8^0 = 1$

$7^1 = 7$

$4^{-1} = \dfrac{1}{4}$

Tip Remember $a^0 = 1$, not zero, as many people seem to think.

For more help see Intermediate Success Guide pages **22** and **23**

Multiplying out brackets

When working on different areas of algebra, you need to be able to multiply out both single and double brackets.

SINGLE BRACKETS

Remember that the item outside the brackets multiplies **each separate item** inside the brackets.

EXAMPLES

$2(a + 3) = 2 \times a + 2 \times 3$
$\qquad = 2a + 6$

$b(3b + 4a) = 3b \times b + 4a \times b$
$\qquad = 3b^2 + 4ab$

$-2(c - 2d) = -2 \times c - (-2 \times 2d)$
$\qquad = -2c + 4d$

Tip You need to take extra care with negative numbers.
Remember:

$+ \times + = +$
$+ \times - = -$
$- \times + = -$
$- \times - = +$

TWO PAIRS OF BRACKETS

Each term in the first pair of brackets is multiplied with each term in the second pair of brackets.

$(a + 3)(a + 2) = a(a + 2) + 3(a + 2)$
$\qquad\qquad = a^2 + 2a + 3a + 6$
$\qquad\qquad = a^2 + 5a + 6$

Tip

Remember to simplify!

$(5a + 3)(a - 2) = 5a^2 - 10a + 3a - 6$
$\qquad\qquad\qquad\quad ① \quad\ ② \quad\ ③ \quad\ ④$
$\qquad\qquad = 5a^2 - 7a - 6$

For more help see **Intermediate Success Guide** page **30**

Factorisation

This process is the reverse of expanding brackets.

It is important to take out the **highest common factor** when factorising single brackets.

SINGLE BRACKETS

EXAMPLES

$12b^2 - 6b = 6b(2b - 1)$

> $6b$ is the highest common factor of $12b^2$ and $6b$.

$4y^3 - 8y^2 = 4y^2(y - 2)$

TWO PAIRS OF BRACKETS

Factorising a quadratic expression of the form $ax^2 + bx + c$ gives two pairs of brackets.

EXAMPLES

$x^2 + 4x + 4 = (x + 2)(x + 2)$
$x^2 - 6x + 5 = (x - 5)(x - 1)$
$x^2 - 2x - 8 = (x - 4)(x + 2)$
$x^2 - 16 = (x - 4)(x + 4)$

> This is known as the difference of two squares.
> $x^2 - a^2 = (x - a)(x + a)$

For more help see Intermediate Success Guide page **31**

Rearranging formulae

You need to know how to rearrange a formula to give it a new subject.

EXAMPLE 1

Make v the subject of this formula.

$u^2 = p + 2vs$

Subtract p from both sides.

$u^2 - p = 2vs$

$\dfrac{u^2 - p}{2s} = v$ or $v = \dfrac{u^2 - p}{2s}$ as it is more commonly written.

Divide both sides by $2s$.

Tip It really is sensible to do these step by step so that you don't go wrong.

EXAMPLE 2

Make a the subject of this formula.

$b = \dfrac{4}{3}(a - 10)$

$b = \dfrac{4(a - 10)}{3}$

Rewritten, the formula is clearer.

$3b = 4(a - 10)$

Multiply both sides by 3.

$\dfrac{3b}{4} = (a - 10)$

Divide both sides by 4.

$\dfrac{3b}{4} + 10 = a$

Add 10 to both sides.

or $\quad a = \dfrac{3b}{4} + 10$

Tip Remember, the expression here is being divided by 3 so it is useful to rewrite it more clearly.

For more help see Intermediate Success Guide page 30

Finding the equation of a straight line

The general equation of a straight line is $y = mx + c$
where m is the **gradient** and c is the **intercept** on the y-axis.

EXAMPLE

Find the equation of this line.

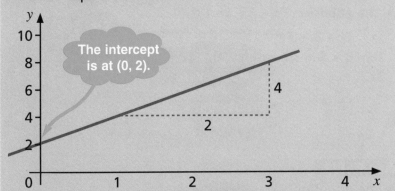

The intercept is at (0, 2).

Find the gradient by drawing in a triangle.
Work out the height and base of the triangle.

$$\text{Gradient} = \frac{\text{height}}{\text{base}}$$

Notice the different scales on the axes! Lots of students just count squares. Don't! It will give a wrong answer, unless the scales are the same.

The gradient is $\frac{4}{2} = 2$.

The equation is $y = 2x + 2$.

> Most students find algebra difficult. Practise lots of questions and see pages 28–43 of the GCSE Success Guide, Intermediate level. The better you are at algebra the more likely it is you will get a grade B.

INTERMEDIATE

For more help see **Intermediate Success Guide** pages **28** to **43**

21

Solving equations

An **algebraic equation** is two algebraic expressions separated by an equals sign. The equals sign is like a balance. When solving equations, whatever you do to one side, you must also do to the other.

ALTERNATIVE METHOD

EXAMPLE 1

This equation has the unknown on both sides.

Solve the equation $5x - 3 = 2x + 8$.

$$5x = 2x + 8 + 3$$

Add 3 to both sides.

$$5x - 2x = 8 + 3$$

Subtract $2x$ from both sides.

$$3x = 11$$

$$x = \frac{11}{3}$$

Divide both sides by 3.

$$x = 3\frac{2}{3}$$

EXAMPLE 2

Solve the equation
$5(3x - 2) = 30$.

Multiply out the brackets.

$$15x - 10 = 30$$

Add 10 to both sides.

$$15x = 30 + 10$$

$$15x = 40$$

$$x = \frac{40}{15}$$

Divide both sides by 15.

$$x = 2\frac{2}{3}$$

ALTERNATIVE METHOD

$$5(3x - 2) = 30$$

$$(3x - 2) = 6$$

Divide both sides by 5.

$$3x = 6 + 2$$

Add 2 to both sides.

$$3x = 8$$

$$x = \frac{8}{3}$$

Divide both sides by 3.

$$x = 2\frac{2}{3}$$

For more help see Intermediate Success Guide page **32**

Solving inequalities

Inequalities

You can solve inequalities in exactly the same way as you would equations, except that when you multiply or divide by a negative number you must reverse the inequality sign.

EXAMPLE 1

Solve $3n + 5 < 12$.

> Subtract 5 from both sides.

$$3n + 5 < 12$$
$$3n < 12 - 5$$

> Divide both sides by 3.

$$3n < 7$$
$$n < \frac{7}{3}$$

EXAMPLE 2

Solve $3n < 7n + 6$.

> Subtract $7n$ from both sides.

$$3n - 7n < 6$$
$$-4n < 6$$

> Divide both sides by −4 and reverse the inequality sign.

$$n > -\frac{6}{4}$$
$$n > -\frac{3}{2}$$

EXAMPLE 3

Solve $6 \leqslant 5n + 4 \leqslant 10$.

This inequality actually represents the two separate inequalities.

$$6 \leqslant 5n + 4 \text{ and } 5n + 4 \leqslant 10$$

$$6 \leqslant 5n + 4 \leqslant 10$$

> Subtract 4 from both sides, in both inequalities.

$$2 \leqslant 5n \leqslant 6$$

> Divide both sides by 5.

$$\frac{2}{5} \leqslant n \leqslant \frac{6}{5}$$

For more help see **Intermediate Success Guide** page **36**

Pythagoras' theorem

Pythagoras' theorem states that, in any right-angled triangle, the square on the hypotenuse is equal to the sum of the squares on the other two sides.

So, in the diagram:

$c^2 = a^2 + b^2$

This can be rearranged to give $a^2 = c^2 - b^2$ or $b^2 = c^2 - a^2$.

The main problems occur when you are asked to find one of the shorter sides of a right-angled triangle.

EXAMPLE

Find the length of the base, AC, of this isosceles triangle.

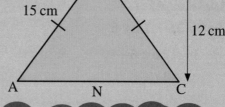

To use Pythagoras' theorem, you need a right-angled triangle. Draw in the perpendicular height, to split the triangle down the middle.

Now using Pythagoras' theorem gives:

$15^2 = AN^2 + 12^2$

$AN^2 = 15^2 - 12^2$

$AN^2 = 225 - 144$

$AN^2 = 81$

$AN = \sqrt{81}$

$AN = 9$ cm

Length of AC = $2 \times 9 = 18$ cm

For more help see **Intermediate Success Guide** pages **60** and **61**

Trigonometry

The three trigonometric ratios are:

$$\sin\theta = \frac{\text{opposite}}{\text{hypotenuse}} \qquad \cos\theta = \frac{\text{adjacent}}{\text{hypotenuse}} \qquad \tan\theta = \frac{\text{opposite}}{\text{adjacent}}$$

The mnemonic **SOH CAH TOA** gives a quick way of remembering the ratios. It comes from the first letters of 'sin = opposite ÷ hypotenuse' and so on.

To key sin 30° into a calculator you may have to enter it backwards, for example, **3 0 SIN**. Check how your calculator does it.

Questions where you need to find the angle cause quite a few problems.

EXAMPLE

Calculate angle CDE.

$$\cos x = \frac{\text{adjacent}}{\text{hypotenuse}}$$

Label the sides and choose the ratio.

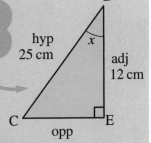

$$\cos x = \frac{12}{25}$$

Divide the top value by the bottom value.

$$\cos x = 0.48$$

Use the second function on your calculator to find angle x.

$$x = \cos^{-1} 0.48$$
$$= 61.3° \ (1 \text{ d.p.})$$

Tip
Always check that the answer sounds sensible. An angle of 0.48° is very small!

For more help see Intermediate Success Guide pages **62**

Similarity

When shapes or solids are **similar**, corresponding lengths are in the same ratio and corresponding angles are equal. You can use this information to find missing lengths.

EXAMPLE

Find the length of AD in these similar triangles.

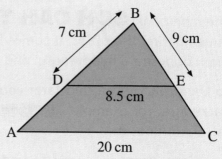

7 cm

9 cm

D

E

8.5 cm

A

C

20 cm

not to scale

Drawing the separate triangles might help.

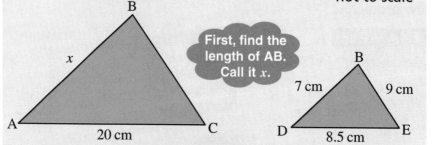

B

x

A

20 cm

C

First, find the length of AB. Call it x.

B

7 cm

9 cm

D

8.5 cm

E

$$\frac{x}{7} = \frac{20}{8.5}$$

Corresponding sides are in the same ratio.

$$x = \frac{20}{8.5} \times 7$$

$x = 16.47$ cm (correct to 2 d.p.)

The length of AD = $16.47 - 7$

$\qquad\qquad\qquad = 9.47$ cm (correct to 2 d.p.)

For more help see Intermediate Success Guide page **68**

Loci

The **locus** of a point is the set of all the possible positions which that point can occupy, subject to some given condition or rule.

> Locus is the singular noun. The plural is **loci**.

Common loci

- The locus of points that are a constant distance from a fixed point is a circle.

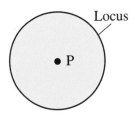

Locus

• P

- The locus of points that are equidistant from two non-parallel lines is the bisector of the angle between the lines.

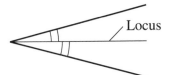

Locus

- The locus of points that are equidistant from two points X and Y is the perpendicular bisector of the line XY.

> **Tip** When answering questions about loci, make accurate constructions, using a ruler and a pair of compasses.

- The locus of points that are at a constant distance from a line XY is a pair of parallel lines, one on either side of XY.
  If the ends of the line are fixed, the locus follows a semicircle around each end.

Perpendicular bisector

X _____ Y

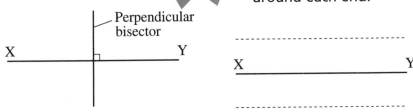

X _____ Y

For more help see **Intermediate Success Guide** page **69**

Averages of grouped data

MEAN

For data grouped into **class intervals,** the mean can only be estimated. Use the midpoint of each class interval.

EXAMPLE

The table shows the heights of some pupils in Year 9.

Height (h cm)	Frequency (f)	Midpoint (x)	$f \times x$
$140 \leqslant h < 145$	4	142.5	570
$145 \leqslant h < 150$	7	147.5	1032.5
$150 \leqslant h < 155$	14	152.5	2135
$155 \leqslant h < 160$	5	157.5	787.5
$160 \leqslant h < 165$	2	162.5	325

Mean $(\bar{x}) = \dfrac{\Sigma fx}{\Sigma f}$

$$= \frac{(4 \times 142.5) + (7 \times 147.5) + (14 \times 152.5) + (5 \times 157.5) + (2 \times 162.5)}{4 + 7 + 14 + 5 + 2}$$

$$= \frac{4850}{32}$$

$$= 151.6 \text{ cm (correct to 1 d.p.)}$$

MODAL CLASS

For grouped data you cannot find the mode, so use the modal class. The modal class is the group with the highest frequency, in this example it is $150 \leqslant h < 155$.

Tip

Remember to divide by the sum of the frequencies.

MEDIAN

You need to find the class interval that includes the median value. There are 32 people in the survey, the middle height is between the sixteenth and seventeenth heights. Both are in the class interval $150 \leqslant h < 155$.

For more help see Intermediate Success Guide page **82**

Moving averages

Moving averages smooth out the changes in a set of data values that vary over a period of time.

A three-point moving average uses three consecutive data items in each calculation. A four-point moving average uses four consecutive data items in each calculation and so on.

EXAMPLE 1

Find the three-point moving averages for the following set of data.

2, 4, 6, 6, 1, 5

Average for first three data points = $(2 + 4 + 6) \div 3 = 4$
Average for next three data points = $(4 + 6 + 6) \div 3 = 5.\dot{3}$
Average for next three data points = $(6 + 6 + 1) \div 3 = 4.\dot{3}$
Average for last three data points = $(6 + 1 + 5) \div 3 = 4$

EXAMPLE 2

Find the four-point moving averages for the following set of data.

4, 6, 2, 5, 4, 3, 4, 5

Average for first four data points = $(4 + 6 + 2 + 5) \div 4 = 4.25$
Average for next four data points = $(6 + 2 + 5 + 4) \div 4 = 4.25$
Average for next four data points = $(2 + 5 + 4 + 3) \div 4 = 3.5$
Average for next four data points = $(5 + 4 + 3 + 4) \div 4 = 4$
Average for last four data points = $(4 + 3 + 4 + 5) \div 4 = 4$

For more help see Intermediate Success Guide page **81**

INTERMEDIATE

29

Cumulative frequency graphs

You can use cumulative frequency graphs to find the **median** and **spread** of grouped data.

The table shows the times, in minutes, taken by 49 pupils to travel to school.

Time (t minutes)	Frequency	Cumulative frequency
$0 \leqslant t < 10$	15	15
$10 \leqslant t < 20$	16	31 (15 + 16)
$20 \leqslant t < 30$	9	40 (31 + 9)
$30 \leqslant t < 40$	6	46 (40 + 6)
$40 \leqslant t < 50$	3	49 (46 + 3)

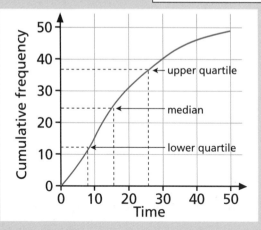

Plot the cumulative frequency diagram

- To plot the points, use the upper class limits of the intervals. In this example, plot (10, 15), (20, 31), (30, 40), (40, 46), (50, 49).
- Since nobody took less than zero time, the graph starts at (0, 0).
- Join the points with a smooth curve.

Tip Use this cumulative frequency to find the median by reading across the vertical scale at 24.5 to the curve and then read down to the horizontal axis to find the time.

For more help see Intermediate Success Guide pages **84** and **85**

Median and interquartile range

The median of the data is the halfway value. So the median time is the time taken by the middle pupil.

The middle pupil is at $\frac{1}{2} \times$ total cumulative frequency $= \frac{1}{2} \times 49 = 24.5$.

The median time is 16 minutes.

The **range** of a set of data tells you how spread out the data values are. The range is affected by extreme values, and so it is useful to consider only the middle 50% of a distribution. This involves finding the **interquartile range**.

Divide the data into four parts or **quartiles** and work out the difference between the upper and lower quartiles.

Lower quartile: $\frac{1}{4} \times 49 = 12.25 \longrightarrow 8$ Read across from 12.25 on the vertical axis.

Upper quartile: $\frac{3}{4} \times 49 = 36.75 \longrightarrow 25.8$ Read across from 36.75 on the vertical axis.

Interquartile range = 25.8 – 8 = 17.8

BOX AND WHISKER DIAGRAMS (BOX PLOT)

These diagrams show the interquartile range as a box. The end values are shown at the ends of the 'whiskers'. They are useful for comparing distributions.

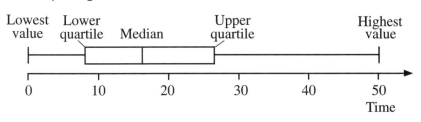

For more help see Intermediate Success Guide pages **84** and **85**

Proportion

GCSE questions on this topic usually involve statements such as:

- a is proportional to the square of b
- c is proportional to the square root of d
- p is inversely proportional to a^2
- d varies as the square of x.

EXAMPLE

a is inversely proportional to the square of b and $a = 16$ when $b = 2$.

a) Write down a formula connecting a and b.

b) Work out the value of b if $a = 25$.

a) $\quad a \propto \dfrac{1}{b^2}$ Write out the statement, using the proportionality symbol.

$\quad a = \dfrac{k}{b^2}$ Replace $\propto$ with $= k$ to make an equation.

$\quad 16 = \dfrac{k}{2^2}$ Substitute the values given in the question, in order to find k.

$16 \times 4 = k$

$\quad k = 64$ Put the value of k back into the equation.

$\quad a = \dfrac{64}{b^2}$

b) If $a = 25$:

$\quad b^2 = \dfrac{64}{25}$ Substitute the value of a into the equation.

$\quad b = \pm\dfrac{8}{5}$ Take the square root to find b.

For more help see **Higher Success Guide** page **26**

Surds

Surds are numbers that include a square root sign, such as $\sqrt{2}$ or $\sqrt{7}$. They are **irrational numbers**.

Remember these rules.

1 $\sqrt{a} \times \sqrt{b} = \sqrt{ab}$ $\qquad$ $\sqrt{5} \times \sqrt{7} = \sqrt{35}$

2 $(\sqrt{b})^2 = \sqrt{b} \times \sqrt{b} = b$ $\quad$ $(\sqrt{7})^2 = \sqrt{7} \times \sqrt{7} = 7$

3 $\dfrac{\sqrt{a}}{\sqrt{b}} = \sqrt{\dfrac{a}{b}}$ $\qquad$ $\dfrac{\sqrt{15}}{\sqrt{3}} = \sqrt{\dfrac{15}{3}} = \sqrt{5}$

4 $(a + \sqrt{b})^2 = (a + \sqrt{b})(a + \sqrt{b}) = a^2 + 2a\sqrt{b} + b$

5 $(a - \sqrt{b})^2 = (a - \sqrt{b})(a - \sqrt{b}) = a^2 - 2a\sqrt{b} + b$

6 $(a + \sqrt{b})(a - \sqrt{b}) = a^2 - b$

EXAMPLE

$$(\sqrt{3} - 2)^2 = (\sqrt{3} - 2)(\sqrt{3} - 2)$$
$$= \sqrt{9} - 4\sqrt{3} + 4$$
$$= 3 - 4\sqrt{3} + 4$$
$$= 7 - 4\sqrt{3}$$

RATIONALISING THE DENOMINATOR

EXAMPLE

Simplify $\dfrac{2}{\sqrt{3}}$.

$$\dfrac{2}{\sqrt{3}} \times \dfrac{\sqrt{3}}{\sqrt{3}} = \dfrac{2\sqrt{3}}{3}$$

Multiply both the numerator and the denominator by $\sqrt{3}$.

For more help see Higher Success Guide page **27**

Indices

In addition to the laws of indices listed on page 17, you also need to know how to work with **negative powers**.

Remember: $a^{-n} = \dfrac{1}{a^n}$

EXAMPLES

$3^{-2} = \dfrac{1}{3^2} = \dfrac{1}{9}$

$5a^{-2} = \dfrac{5}{a^2}$

$4^{-3} = \dfrac{1}{4^3} = \dfrac{1}{64}$

POWERS WITH FRACTIONS

The laws of indices

$$a^{\frac{1}{m}} = \sqrt[m]{a} \qquad a^{\frac{n}{m}} = (\sqrt[m]{a})^n$$

Tip Remember, you must find the root of the number first and then raise this root to the given power.

EXAMPLES

$36^{\frac{1}{2}} = \sqrt[2]{36} = \pm 6$

$27^{\frac{1}{3}} = \sqrt[3]{27} = 3$

$32^{\frac{1}{5}} = \sqrt[5]{32} = 2$

$8^{\frac{2}{3}} = (\sqrt[3]{8})^2 = 2^2 = 4$

When you mean the square root, you usually just write $\sqrt{\ }$, not $\sqrt[2]{\ }$.

For more help see **Higher Success Guide** page **23**

Upper and lower bounds

If a weight, w, is given as **15 grams to the nearest gram**, the actual weight is in the interval $14.5\,g \leqslant w \leqslant 15.4999\ldots\,g$.

In this context, the weight 15.4999... grams is usually written as 15.5 grams, although it is accepted that 15.5 grams would be rounded to 16 grams (to the nearest gram).

14.5 g is called the **lower bound**.
15.5 g is called the **upper bound**.
$14.5 \leqslant w < 15.5$

EXAMPLE

To the nearest whole number, the value of a is 207 and the value of b is 12. Calculate the maximum and minimum values of these expressions.

a) $a + b$ **b)** $a - b$ **c)** ab **d)** $\dfrac{a}{b}$

Maximum value of a = 207.5   Maximum value of b = 12.5
Minimum value of a = 206.5   Minimum value of b = 11.5

a) $a + b$
Maximum = 207.5 + 12.5 = 220
Minimum = 206.5 + 11.5 = 218

b) $a - b$
Maximum = 207.5 − 11.5 = 196
Minimum = 206.5 − 12.5 = 194

c) ab
Maximum = 207.5 × 12.5 = 2593.75
Minimum = 206.5 × 11.5 = 2374.75

d) $\dfrac{a}{b}$
Maximum = 207.5 ÷ 11.5 = 18.04 (correct to 2 d.p.)
Minimum = 206.5 ÷ 12.5 = 16.52

Tip Take great care with calculations involving subtraction and division.

For more help see Higher Success Guide page **28**

Solving quadratic equations

When a quadratic equation cannot be solved by factorisation alternative methods are needed.

USING THE FORMULA

You can use the formula $x = \dfrac{-b \pm \sqrt{b^2 - 4ac}}{2a}$

to solve quadratic equations of the form $ax^2 + bx + c = 0$, $a \pm 0$.

EXAMPLE

Solve the equation $2x^2 - 4x - 5 = 0$, giving your answer correct to two decimal places.

Comparing the quadratic equation with the general equation $ax^2 + bx + c$, note that $a = 2$, $b = -4$, $c = -5$. Substituting these values in the formula:

$$x = \frac{-b \pm \sqrt{b^2 - 4ac}}{2a}$$

$$x = \frac{-(-4) \pm \sqrt{(-4)^2 - (4 \times 2 \times -5)}}{2 \times 2}$$

$$= \frac{4 \pm \sqrt{16 - (-40)}}{4}$$

$$x = \frac{4 \pm \sqrt{56}}{4}$$

Tip
Always check that the quadratic equation is written with the right-hand side equal to zero.

So $x = \dfrac{4 + \sqrt{56}}{4}$ or $x = \dfrac{4 - \sqrt{56}}{4}$

$x = 2.87$ or $x = -0.87$ (correct to 2 d.p.)

For more help see Higher Success Guide page **40**

Solving quadratic equations

COMPLETING THE SQUARE

To do this, you need to express the quadratic equation in the form $(x + a)^2 + b = 0$.

EXAMPLE

Express $x^2 - 6x + 3 = 0$ as a completed square and hence solve it.

Rearrange the equation in the form $ax^2 + bx + c = 0$ and check that $a = 1$.

$x^2 - 6x + 3 = 0$ is in the required form.

Write an expression in the form $(x + \frac{b}{2})^2$.

$(x - 3)^2$

Multiply out the brackets and compare the expression to $x^2 - 6x + 3$.
$(x - 3)(x - 3) = x^2 - 6x + 9$

To make $x^2 - 6x + 9$ like the original, subtract 6.
$(x - 3)^2 - 6 = 0$

Now solve the equation.
$$(x - 3)^2 = 6$$
$$x - 3 = \pm \sqrt{6}$$

Therefore $x = \sqrt{6} + 3$ or $x = -\sqrt{6} + 3$
$$x = 5.45 \qquad x = 0.55 \text{ (correct to 2 d.p.)}$$

For more help see **Higher Success Guide** page **40**

Algebraic fractions

At the Higher level you are required to simplify complicated expressions. To do this you will need to factorise parts of the expression and then cancel if appropriate.

SIMPLIFYING

EXAMPLE 1

Simplify

$$\frac{4n + 8}{n^2 - 4n + 3} \times \frac{n^2 - 7n + 12}{n^2 - 4}.$$

$$\frac{4n + 8}{n^2 - 4n + 3} \times \frac{n^2 - 7n + 12}{n^2 - 4}.$$

$$= \frac{4(n + 2)}{(n - 1)(n - 3)} \times \frac{(n - 3)(n - 4)}{(n - 2)(n + 2)}$$
First, factorise each part of each fraction.

$$= \frac{4\cancel{(n + 2)}}{(n - 1)\cancel{(n - 3)}} \times \frac{\cancel{(n - 3)}(n - 4)}{(n - 2)\cancel{(n + 2)}}$$
Next, cancel common factors.

$$= \frac{4(n - 4)}{(n - 1)(n - 2)}$$
Finally, rewrite the answer in its simplest form.

EXAMPLE 2

Simplify fully the expression $\dfrac{n^2 - 9}{2n^2 - 7n + 3}$.

> Remember the difference of two squares:
> $(x^2 - a^2)$
> $= (x - a)(x + a)$

$n^2 - 9 = (n - 3)(n + 3)$ Factorise the numerator.
$2n^2 - 7n + 3 = (2n - 1)(n - 3)$ Factorise the denominator.

$$\frac{n^2 - 9}{2n^2 - 7n + 3} = \frac{(n + 3)(n - 3)}{(2n - 1)(n - 3)}$$
Combine these.

$$= \frac{(n + 3)\cancel{(n - 3)}}{(2n - 1)\cancel{(n - 3)}}$$
Cancel the common factors.

$$= \frac{n + 3}{2n - 1}$$

For more help see **Higher Success Guide** page **41**

Solving algebraic fractions

Solving algebraic fractions can be difficult, as it can involve several steps. You may need to solve quadratic equations by **factorising** or by **using the formula**.

When combining algebraic fractions, always try to choose the **lowest common denominator**. For example, for

$$\frac{1}{(x-2)(x+1)} + \frac{2}{(x-2)}$$

the lowest common denominator is $(x-2)(x+1)$, not $(x-2)^2(x+1)$.

EXAMPLE

Solve the equation $\dfrac{1}{(2n+1)} - \dfrac{1}{(n-4)} = 1$.

Write down the lowest common denominator and adjust the numerators.

$$\frac{(n-4) - (2n+1)}{(2n+1)(n-4)} = 1$$

$$(n-4) - (2n+1) = 1(2n+1)(n-4)$$

Multiply both sides by $(2n+1)(n-4)$

$$-n - 5 = 2n^2 - 7n - 4$$
$$2n^2 - 7n - 4 + n + 5 = 0$$
$$2n^2 - 6n + 1 = 0$$

Simplify.

$$n = \frac{6 \pm \sqrt{36 - (4 \times 2 \times 1)}}{4}$$

$$= \frac{6 \pm \sqrt{28}}{4}$$

$n = 2.82$ or $n = 0.18$
(correct to 2 d. p.)

Solve, using the quadratic formula.
$$x = \frac{-b \pm \sqrt{b^2 - 4ac}}{2a} \text{ where}$$
$x = n, a = 2, b = -6, c = 1$

For more help see **Higher Success Guide** pages **40** and **41**

Rearranging formulae

This involves rewriting the formula to change the subject. When the subject appears in more than one term it becomes more difficult.

EXAMPLE 1

Make a the subject of the formula $ab = ac + b$.

$$ab = ac + b$$

$ab - ac = b$ Collect the terms involving a on one side of the equation.

$a(b - c) = b$ Factorise.

$$a = \frac{b}{b - c}$$ Divide both sides by $(b - c)$.

EXAMPLE 2

Make s the subject of the formula $p = \dfrac{s + a}{s - b}$.

$$p = \frac{s + a}{s - b}$$

$p(s - b) = s + a$ Multiply both sides by $(s - b)$.

$ps - pb = s + a$ Multiply out the brackets.

$ps - s = a + pb$ Collect the terms involving s on one side of the equation.

$s(p - 1) = a + pb$ Factorise.

$$s = \frac{a + pb}{p - 1}$$ Rewrite in its simplest form.

For more help see **Higher Success Guide** page **41**

A circle and a line

You may be asked to find coordinates of points where a circle cuts a straight line. From Pythagoras' theorem, the equation of a circle, centre (0, 0) and radius r, is $x^2 + y^2 = r^2$.

EXAMPLE 1

Find the coordinates of the points where the line $y - 5 = x$ cuts the circle $x^2 + y^2 = 25$.

> The points of intersection must lie on the line and on the circle. The coordinates must satisfy both equations.

$y - 5 = x$	**1**
$x^2 + y^2 = 25$	**2**

$y = x + 5$ **3**   Rewrite equation **1** in the form '$y = ...$'.

$x^2 + (x + 5)^2 = 25$ Eliminate y by substituting **3** into **2**.

$x^2 + x^2 + 10x + 25 = 25$ Remember: $(x + 5)^2 = (x + 5)(x + 5)$.

$2x^2 + 10x = 0$ Simplify.

$2x(x + 5) = 0$ Factorise.

So $x = 0$ or $x = -5$.

If $x = 0$ $y = 0 + 5 = 5$ Substitute the values of x into equation **3** to find the values of y.

If $x = -5$ $y = -5 + 5 = 0$

The line $y - 5 = x$ cuts the circle $x^2 + y^2 = 25$ at (0, 5) and (5, 0).

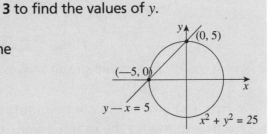

EXAMPLE 2

Try solving these simultaneous equations.

$y = x + 5$	**1**
$x^2 + y^2 = 13$	**2**

You should get the answer $x = -2$, $y = 3$ and $x = -3$, $y = 2$.

For more help see **Higher Success Guide** page **48**

Functions and transformations

At the Higher level you need to know and use **function notation**. You also need to be able to **transform graphs**.

Function notation is used to describe the relationship between two variables such as x and y.

In the relationship $y = ...$ the right-hand side may be written as $f(x)$, giving $y = f(x)$. This is read as 'y is a function of x.'

There are four different graph transformations which you need to know.

1 $y = f(x) \pm a$ The graph moves up or down the y-axis by a value of a.

> This graph has moved 2 units up the y-axis.

$y = f(x) + 2$
$y = f(x)$
$y = f(x) - 2$

> This graph has moved 2 units down the y-axis.

2 $y = f(x \pm a)$ The whole graph moves a units to the left or right.

$y = f(x + a)$ moves the graph a units to the left.
$y = f(x - a)$ moves the graph a units to the right.

> This graph has moved 2 units to the left.

$y = f(x)$
$y = f(x + 2)$
$y = f(x - 2)$

> This graph has moved 2 units to the right.

3 $y = k \times \text{f}(x)$ The original graph stretches (or shrinks if $k < 1$) along the y-axis by a factor of k.

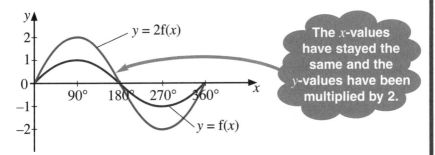

The x-values have stayed the same and the y-values have been multiplied by 2.

4 $y = \text{f}(kx)$ The original graph stretches inwards in the x-direction by $\frac{1}{k}$, if $k > 1$.

If $k < 1$ then the graph stretches outwards in the x-direction by $\frac{1}{k}$.

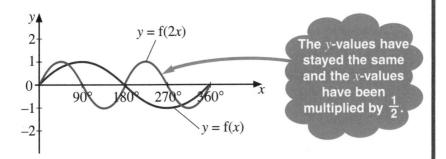

The y-values have stayed the same and the x-values have been multiplied by $\frac{1}{2}$.

For more help see **Higher Success Guide** page **49**

Trigonometry in three dimensions

When you are completing questions involving three-dimensional figures, it is essential that you identify the right-angled triangles you are using. Then you can apply trigonometry and Pythagoras' theorem.

EXAMPLE

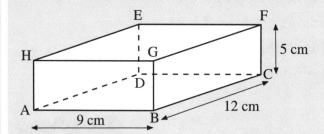

The diagram shows a cuboid. Work out:

a) the length of AC
b) the size of angle FAC
c) the size of angle HBA.

a) Using triangle ABC:

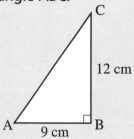

$AC^2 = 12^2 + 9^2 = 144 + 81$ By Pythagoras' theorem.

$AC^2 = 225$

$AC = \sqrt{225}$

$AC = 15$ cm

b) Using triangle ACF:

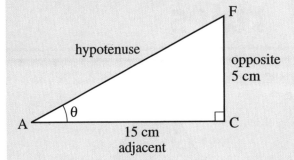

$$\tan \theta = \frac{\text{opposite}}{\text{adjacent}}$$

$$\tan \theta = \frac{5}{15}$$

$\theta = \tan^{-1} 0.3333\ldots$

$\theta = 18.4°$ (correct to 1 d.p.)

θ is a Greek letter, theta, often used for angles.

c) Using triangle ABH:

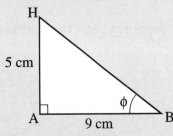

$$\tan \phi = \frac{\text{opposite}}{\text{adjacent}}$$

$$\tan \phi = \frac{5}{9}$$

$\phi = \tan^{-1} 0.5555\ldots$

$\phi = 29.05°$

$\phi = 29.1°$ (correct to 1 d.p.)

ϕ is another Greek letter, phi, often used for angles.

For more help see Higher Success Guide page **72**

The sine and cosine rules

You can use the sine rule and the cosine rule to solve problems about triangles that do not contain right angles.

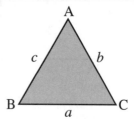

THE SINE RULE

$$\frac{a}{\sin A} = \frac{b}{\sin B} = \frac{c}{\sin C} \quad \text{or} \quad \frac{\sin A}{a} = \frac{\sin B}{b} = \frac{\sin C}{c}$$

Use the sine rule when you are working with **two sides and two angles**.

THE COSINE RULE

$$a^2 = b^2 + c^2 - (2bc \cos A) \quad \text{or} \quad \cos A = \frac{b^2 + c^2 - a^2}{2bc}$$

Use the cosine rule when you are working with **three sides and one angle**.

EXAMPLE 1

Find the size of angle PQR.

Call angle PQR θ.

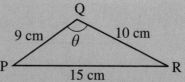

You are given information about three sides and one angle, so use the cosine rule.

$$\cos \theta = \frac{9^2 + 10^2 - 15^2}{2 \times 9 \times 10} = \frac{81 + 100 - 225}{180} = \frac{-44}{180}$$

$\cos \theta = -0.2\dot{4}$

$\theta = \cos^{-1} -0.2\dot{4}$

$\theta = 104.15°$

$\quad = 104°$ (to the nearest degree)

> The negative sign shows that the angle is obtuse.

EXAMPLE 2

Find:

a) the area of triangle ABC

b) the length of AB.

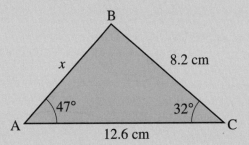

a) To find the area of triangle ABC you need two sides and the included angle.

$$\text{Area} = \tfrac{1}{2} \times a \times b \times \sin C$$
$$= \tfrac{1}{2} \times 12.6 \times 8.2 \times \sin 32°$$

$$\text{Area} = 27.4 \text{ cm}^2$$

b) To find the length of AB, use the sine rule.
Call the length of AB, which you need to find, x.

$$\frac{x}{\sin 32°} = \frac{8.2}{\sin 47°}$$
$$x = \frac{8.2}{\sin 47°} \times \sin 32°$$
$$x = 5.94 \text{ cm}$$

Tip Remember to use the form of the sine rule that will give you the missing length as the numerator of one of the fractions.

For more help see **Higher Success Guide** page **74**

Arc length

Make sure that you remember the formula.

Arc length = $\dfrac{\theta}{360°} \times 2\pi r$

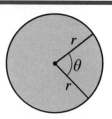

Find the perimeter of this sector.

The length of the arc

$$= \dfrac{55°}{360°} \times 2 \times \pi \times 12$$

$$= 11.52 \text{ cm (correct to 2 d.p.)}$$

Perimeter = 24 + 11.52

$$= 35.52 \text{ cm (correct to 2 d.p.)}$$

12 cm   12 cm

55°

The length of the arc in this sector is 20 cm. Find the value of θ giving your answer to the nearest degree.

$$\dfrac{\theta}{360°} \times 2 \times \pi \times 8 = 20$$

$$\dfrac{\theta}{360°} \times 16\pi = 20$$

$$\dfrac{\theta}{360°} = \dfrac{20}{16\pi}$$

$$\theta = \dfrac{20}{16\pi} \times 360°$$

$$\theta = 143° \text{ (to the nearest degree)}$$

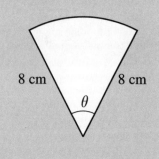

8 cm   8 cm

θ

For more help see **Higher Success Guide** page **82**

Sector area

Make sure that you remember the formula.

Sector area = $\dfrac{\theta}{360°} \times \pi r^2$

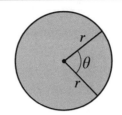

EXAMPLE

Find the area of the shaded region in the diagram.

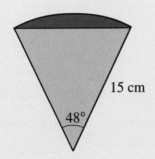

15 cm

48°

Find the area of the sector.

Area = $\dfrac{48°}{360°} \times \pi \times 15^2$

= 94.2477... cm²

Find the area of the triangle.

Area = $\frac{1}{2} \times a \times b \times \sin C$

Area = $\frac{1}{2} \times 15 \times 15 \times \sin 48°$

Area = 83.603 87... cm²

a and *b* are both equal to the radius of the sector.

Shaded segment = area of sector − area of triangle

= 94.2477... − 83.6037...

= 10.64 cm²

(correct to 2 d.p.)

Notice that the values have not been rounded until the end.

For more help see **Higher Success Guide** page **82**

Vector geometry

This is a topic that causes many students problems, and is usually on the examination paper.

EXAMPLE

ABCDEF is a regular hexagon.

$\overrightarrow{OB} = $ **a** and $\overrightarrow{OC} = $ **b**.

a) Find, in terms of **a** and **b**, the vectors:
   (i) $\overrightarrow{BC}$ **(ii)** $\overrightarrow{AD}$.

b) Write down the vector $\overrightarrow{FE}$.

c) What geometrical property is shown by the vectors $\overrightarrow{FE}$ and $\overrightarrow{AD}$?

a) **(i)** $\overrightarrow{BC} = -$**a** $+$ **b**
    (ii) $\overrightarrow{AD} = -2$**a** $+ 2$**b**

Remember the triangle rule for vectors.

b) $\overrightarrow{FE} = -$**a** $+$ **b**   Since $\overrightarrow{BC}$ and $\overrightarrow{FE}$ are parallel.

c) $\overrightarrow{FE} = -$**a** $+$ **b** and $\overrightarrow{AD} = 2(-$**a** $+$ **b**$) = 2\overrightarrow{FE}$

So vector $\overrightarrow{AD}$ is twice the length (or magnitude) of vector $\overrightarrow{FE}$ and is parallel to it.

For more help see **Higher Success Guide** pages **84** and **85**

Tree diagrams

Use a tree diagram when you need to show the possible outcomes of two or more events.

EXAMPLE

A bag contains four yellow counters, three red and two green. A counter is taken out of the bag at random, its colour is noted and then it is replaced in the bag and a second counter is taken out of the bag. What is the probability of choosing two counters of different colours?

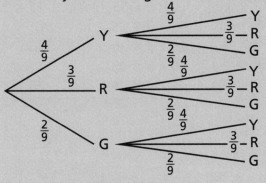

The probabilities for the colours do not change since the first counter is replaced.

The probability of choosing two different colours is:

$P(YR) + P(YG) + P(RY) + P(RG) + P(GY) + P(GR)$

$= (\frac{4}{9} \times \frac{3}{9}) + (\frac{4}{9} \times \frac{2}{9}) + (\frac{3}{9} \times \frac{4}{9}) + (\frac{3}{9} \times \frac{2}{9}) + (\frac{2}{9} \times \frac{4}{9}) \times (\frac{2}{9} \times \frac{3}{9})$

Multiply along the branches.

$= \frac{12}{81} + \frac{8}{81} + \frac{12}{81} + \frac{6}{81} + \frac{8}{81} + \frac{6}{81} = \frac{52}{81}$

Add up the probabilities.

Alternatively, use the fact that the total of all the probabilities must be 1.

The probability of choosing two different colours = 1 – the probability of choosing the same colour each time.

$1 - [P(RR) + P(YY) + P(GG)] = 1 - [(\frac{3}{9} \times \frac{3}{9}) + (\frac{4}{9} \times \frac{4}{9}) + (\frac{2}{9} \times \frac{2}{9})]$

$= 1 - \frac{29}{81} = \frac{52}{81}$

For more help see **Higher Success Guide** page **104**

Histograms

Histograms are like bar charts except that the frequency is represented by the **area of each bar,** rather than the length or height. When the widths of the columns are different the vertical axis represents the **frequency density**.

$$\text{Frequency density} = \frac{\text{frequency}}{\text{class width}}$$

EXAMPLE

The table and histogram give information about the distance, d km, travelled to work by some employees.

Distance (d km)	Frequency
$0 < d \leqslant 15$	15
$15 < d \leqslant 25$	...
$25 < d \leqslant 30$	35
$30 < d \leqslant 45$	...
$45 < d \leqslant 55$	20

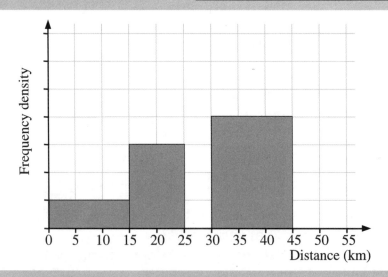

a) Use the information in the histogram to complete the table.

b) Use the table to complete the histogram.

a) To complete the table find the class widths and frequency densities.

Distance (d km)	Class width	Frequency	Frequency density
$0 < d \leqslant 15$	15	15	1
$15 < d \leqslant 25$	10	30	3
$25 < d \leqslant 30$	5	35	7
$30 < d \leqslant 45$	15	60	4
$45 < d \leqslant 55$	10	20	2

Frequency density $= \dfrac{\text{frequency}}{\text{class width}}$

$\quad\quad\quad = \dfrac{15}{15}$

$\quad\quad\quad = 1$

b) Insert a scale on the verticle axis and fill in the missing bars.

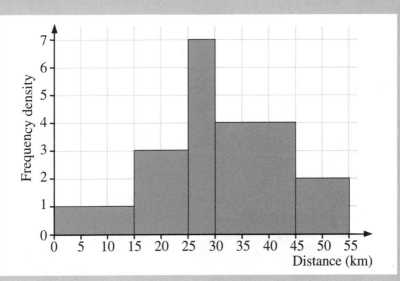

For more help see **Higher Success Guide** page **100**

Multiply and divide

For the non-calculator paper, you need to be able to work competently with numbers. Here are some of the techniques you need to use.

LONG MULTIPLICATION

Multiplying two or more numbers together is known as **finding the product**. It can sometimes help to use a grid method when you are doing long multiplication calculations.

EXAMPLE

Work out 189×23.

Traditional method:

```
   189
 × 23
   567   (189 × 3)
  3780   (189 × 20)
  4347
```

Grid method:

	100	80	9	
20	2000	1600	180	3780
3	300	240	27	+ 567
				4347

LONG DIVISION

When you are carrying out long and short division, take care not to miss out important zeros.

EXAMPLE

Divide 1036 by 74.

```
      14
 74)1036
   - 74
    296
  - 296
      0
```

Always try to estimate the answer to the division first.

or

```
 74 )1036
    -740   74 × 10
     296
   - 296   74 × 4
       0
 Answer
 10 + 4 = 14
```

This method is known as chunking.

MULTIPLYING AND DIVIDING BY NUMBERS BETWEEN 0 AND 1

Multiplying

When multiplying by a number between 0 and 1, remember that the answer is always smaller than the starting value.

EXAMPLES

$5 \times 0.1 = 0.5$ The same as $5 \times \frac{1}{10} = \frac{5}{10} = 0.5$.

$0.3 \times 0.01 = 0.003$ The same as $(3 \times 1) \div 1000$.

$0.42 \times 0.002 = 0.00084$

Tip Always check that the answer is smaller than the starting value.

Dividing

When dividing by a number between 0 and 1, remember that the answer is always bigger than the starting value.

EXAMPLES

$4 \div 0.1 = 40$ The same as $\frac{40}{1}$ (multiply top and bottom by 10).

$42 \div 0.01 = 4200$

$762 \div 0.001 = 762\,000$ The same as $\frac{762\,000}{1}$ (multiply top and bottom by 1000).

Tip Always check that the answer is bigger than the starting value.

Fractions

ADDING AND SUBTRACTING

You can only add or subtract fractions if they have the same denominator.

EXAMPLE 1

$\frac{5}{7} + \frac{11}{28}$

The lowest common denominator of $\frac{1}{7}$ and $\frac{1}{28}$ is 28. Replace $\frac{5}{7}$ with $\frac{20}{28}$.

$\frac{20}{28} + \frac{11}{28} = \frac{31}{28} = 1\frac{3}{28}$

Only add the numerators. The denominator stay the same.

EXAMPLE 2

$\frac{12}{17} - \frac{1}{3}$

The lowest common denominator of $\frac{1}{17}$ and $\frac{1}{3}$ is 51.
Replace $\frac{12}{17}$ with $\frac{36}{51}$ and $\frac{1}{3}$ with $\frac{17}{51}$.

$\frac{36}{51} - \frac{17}{51} = \frac{19}{51}$

EXAMPLE 3

$2\frac{7}{11} + 3\frac{2}{5}$

Add up the whole numbers $2 + 3 = 5$.

$\frac{7}{11} + \frac{2}{5}$

The lowest common denominator of $\frac{1}{11}$ and $\frac{1}{5}$ is 55.
Replace $\frac{7}{11}$ with $\frac{35}{55}$ and $\frac{2}{5}$ with $\frac{22}{55}$.

$\frac{35}{55} + \frac{22}{55} = \frac{57}{55} = 1\frac{2}{55}$

$5 + 1\frac{2}{55} = 6\frac{2}{55}$

Remember to add the 5 from the whole numbers.

So $2\frac{7}{11} + 3\frac{2}{5} = 6\frac{2}{55}$

MULTIPLYING FRACTIONS

To multiply fractions, multiply the numerators together and multiply the denominators together. Always try cancelling before multiplying.

EXAMPLE 1

$\frac{2}{7} \times \frac{4}{11} = \frac{8}{77}$

Tip Write mixed numbers or whole numbers as improper fractions before you start multiplying.

EXAMPLE 2

$3\frac{1}{2} \times \frac{4}{9} = \frac{7}{2} \times \frac{\cancel{4}^{2}}{9}$

$= \frac{14}{9}$

$= 1\frac{5}{9}$

DIVIDING FRACTIONS

To divide one fraction by another, change the division into a multiplication by **taking the reciprocal** of the second fraction, and then multiplying the fractions together.

Tip Taking the reciprocal of a fraction is like turning it upside down.

EXAMPLE

$\frac{7}{9} \div \frac{11}{18}$

$= \frac{7}{9} \times \frac{18}{11}$ Take the reciprocal of $\frac{11}{18}$ and multiply $\frac{7}{9}$ by the result.

$= \frac{7}{9} \times \frac{\cancel{18}^{2}}{11}$ **Try to cancel before multiplying.**

$= \frac{14}{11}$

$= 1\frac{3}{11}$ Write the final answer as a mixed number.

Percentages and estimating

PERCENTAGES

To use a mental method for calculating a percentage of a quantity, find 10% or 1% first.

EXAMPLE

Without using a calculator, find 17.5% of £320.

10% of £320 = 320 ÷ 10 = £32

5% of £320 = £16

2.5% of £320 = £8

17.5% of £320 = 32 + 16 + 8
= £56

ESTIMATING

Estimating is a good way of checking answers. When estimating:

- round the numbers to **one significant figure** (1 s.f.)
- use these easy numbers to work out the estimate
- use the symbol ≈ which means 'is approximately equal to'.

When multiplying or dividing, never approximate a number to zero.

- Use 0.1, 0.01, 0.001, ...

EXAMPLE

Estimate the answer to $\frac{10.2 \times 203}{(3.1)^2}$.

$$\frac{10.2 \times 203}{(3.1)^2} \approx \frac{10 \times 200}{3^2} = \frac{2000}{9} \approx \frac{2000}{10} = 200$$

Standard index form

On the non-calculator paper, when you are multiplying and dividing numbers written in standard form use the **laws of indices**.

EXAMPLE 1

$$(3.2 \times 10^{-6}) \times (4 \times 10^{9}) = (3.2 \times 4) \times (10^{-6} \times 10^{9})$$
$$= 12.8 \times 10^{-6+9}$$
$$= 12.8 \times 10^{3}$$
$$= 1.28 \times 10^{4}$$

Notice that 12.8 is not a number between 1 and 10.

EXAMPLE 2

$$(16.8 \times 10^{9}) \div (4 \times 10^{-3}) = (16.8 \div 4) \times (10^{9} \div 10^{-3})$$
$$= 4.2 \times 10^{9-(-3)}$$
$$= 4.2 \times 10^{12}$$

EXAMPLE 3

$$(6 \times 10^{4})^{2} = (6 \times 10^{4}) \times (6 \times 10^{4})$$
$$= (6 \times 6) \times (10^{4} \times 10^{4})$$
$$= 36 \times 10^{8}$$
$$= 3.6 \times 10^{9}$$

Alternatively:
$$(6 \times 10^{4})^{2} = 6^{2} \times (10^{4})^{2}$$
$$= 36 \times 10^{8}$$
$$= 3.6 \times 10^{9}$$

A close look at AO1: Using and applying mathematics

ASSESSMENT OBJECTIVE 1

(AO1) is known as *Using and applying mathematics* and is used to assess the coursework element of GCSE mathematics. Syllabuses for GCSE vary a little in the way that AO1 is assessed. However, all AO1 work will be assessed using the same criteria and the same scale of marking. The final mark will account for 10% of your final GCSE grade, no matter which syllabus you are following.

In AO1, *Using and applying mathematics*, you are required to use your understanding and knowledge of Number, Algebra and Shape, space and measures to develop your investigation and practical skills. AO1 is used to assess your skills in three key areas:

- **Strand 1 – Making and monitoring decisions:** deciding how to tackle a piece of work and carrying it through to a successful conclusion

- **Strand 2 – Communicating mathematically**: presenting the information in a variety of ways, such as graphs, tables, charts, symbols and algebra, with an explanation as to their use

- **Strand 3 – Developing skills of mathematical reasoning**: making and testing generalisations, justifying and proving why generalisations work and how you alter your approach in the light of the results so far.

Each of these strands is assessed and marks are awarded out of a maximum of 8. If more that one task is submitted, only the best mark in each strand counts towards your overall mark. In marking your coursework, your teacher or examination board uses a set of criteria and measures the quality of your work against them. The table on pages 62 and 63 shows the bases of these criteria.

MARKING

When marking your work, your teacher or the examination board needs to find that statement which best fits your work in each of the three strands. For example, someone who has used diagrams, words and symbols within their work, but has not explained why, would obtain only 3 marks instead of the possible 4 marks in **Strand 2**.

TYPES OF COURSEWORK

Depending on the examination board which your school uses, AO1 is assessed using one of several methods. These are:

1 At least one task which demonstrates ability to use and apply mathematics. The tasks can be practical and/or investigational and are marked by the school.

2 At least one task which demonstrates ability to use and apply mathematics. The tasks can be practical and/or investigational and are marked by the examination board.

No matter which type of assessment you are doing, the same procedures need to be adopted when carrying out investigations.

BEFORE GETTING STARTED

Your teacher will explain which type of coursework you will be doing. The technique for doing investigations and practical work is governed by a general process.

1 Choose the task carefully, read it through and make sure you understand the problem. Make a plan before you start and check it with your teacher. The plan should include:
   - equipment which needs to be used, e.g. calculators, multilink
   - the order in which you are carrying out the problem
   - the data that you need to collect
   - the direction in which you see the investigation going, e.g. the type of questions you need to ask and what extension you might do.

2 Set out a timeplan and establish deadlines by which you must finish particular sections. Don't be tempted to leave your coursework until the last few days as you will rush it and will not produce your best work. Aim to do a small amount every day.

3 Allow yourself plenty of time. In the last few days, ask an adult to read through your work and check that it makes sense.

4 Use levels of mathematics appropriate to your tier of entry to the examination. In particular, proper use of algebra is essential to progress beyond about 4 marks in each of the three strands. For example, writing 'the next term is three times the previous term plus two' would not attract as much credit as $x_{n+1} = 3x_n + 2$.

5 Remember to seek advice from your teacher when necessary.

Once you've started, there are a number of things which you will need to remember in order to improve the quality of your work. See the tips given on the pages that follow the table.

Using and applying mathematics
Assessment criteria

	Strand 1: Making and monitoring decisions
1 mark	I can decide on what I need and start the problem, producing some information or results.
2 marks	I can work through the problem with different examples and I obtain some results.
3 marks	I can obtain further information to help me with the task, and check my results to see if they are sensible.
4 marks	I can work through a complex task by breaking it down into smaller tasks.
5 marks	I can extend the original problem by introducing ideas of my own.
6 marks	I can introduce new methods, e.g. algebra, when working through my task. I can reflect on the work I am doing and explain why I am doing it that way.
7 marks	I can analyse new methods introduced into my work, and give detailed reasons why I am using them. My task is complex, involving at least three features or variables.
8 marks	I have considered and evaluated a number of approaches to a task, and explored a new area of mathematics. I have applied a range of appropriate mathematical techniques.

Strand 2: Communicating mathematically	Strand 3: Developing the skills of mathematical reasoning
I can organise my work and explain what I am thinking.	I can make a simple statement about my results and show an example.
I can clearly present any information or results in tables or diagrams.	I can search for a pattern by looking at my results in a few examples.
I can use tables and diagrams to present my information and can explain patterns I have found, in words and symbols.	I can make rules about the patterns I have found.
I can use and explain a range of tables, charts, diagrams and symbols when presenting my findings.	I can show that the rules I have found work by testing some examples.
I can justify why I have chosen a particular way to present my results.	I can justify the rules I have made by looking at the mathematical structure of the task.
I can explain my results by using mathematical symbols, e.g. algebra.	I can explain how I obtained my rules, and how they help me to progress with the problem.
I can use symbols and mathematical language accurately.	Using several features in my work, I can give a conclusion which is justified by my findings.
I have used symbols, mathematical language and graphical information in a concise and efficient way.	I can give a detailed justification, argument or proof of my solution to a complex problem.

Mathematical investigations

1 Read the question carefully, checking you understand what is required, and introduce the question in your own words.

2 Make a plan of what you are going to do and what materials you need.

3 Decide how you can break the problem down into simple parts. Always keep the problem simple at the start. Illustrate, by example, the processes required.

4 Devise at least four or five simple examples. Draw diagrams to help you collect your results. Decide what to do with your results before you write them down. Explain your choices.

5 Whenever possible, use calculators and computers in your work. If you use a computer, make sure you explain what programs you have used and how you have used them.

6 Use tables and diagrams to record and tabulate your results. Explain why you have chosen them.

7 Make general statements about the rules and patterns you may have found, for instance, 'the numbers go up by 4 each time'. Predict what the next few may be and test with more examples. Try to make use of counter examples.

8 Look at different ways of showing your information and comment on their use. Always ask yourself if you could have shown the information in a different way. If so write it down and compare the different ways. Consider using:

  • tables • diagrams • graphs.

9 Try to find a general rule which, ideally, you can write in symbols. Explain why your rule works. Can you see a connection between the structure and the rule?

10 Try to prove any rules you have made.

11 Once you have finished the original problem, try to decide on alternative ways in which to extend it. Maybe you can change the original structure. However, make sure you explain why you are changing the problem.

Tips for carrying out mathematical investigations

Remember

- *Decide* what to do.
- *Show* how you do it.
- *Explain* why you are doing it.
- *Use* as much mathematics as possible.
- Only make conclusions you have checked carefully.
- *Read* your report through at the end to make sure it makes sense.

Do's and don'ts

You may find these do's and don'ts helpful when you're planning and getting started with coursework.

First, a few things to aim for.

1 **Do** make sure that the coursework is your best work and is clearly and neatly presented.

2 **Do** write a plan of what you intend to do, then check it with your teacher.

3 **Do** label any diagrams, tables, charts and graphs.

4 **Do** explain everything you do, even minor things.

5 **Do** check any answers you obtain and make sure they are sensible.

6 **Do** make sure that you have answered the original problem.

Now, here are a few things to avoid.

1 **Don't** spend a lot of time copying out information which has already been given to you in the question. You may score points for restarting the question in your own words but not for merely repeating it.

2 **Don't** rush your work or leave it until the last minute.

3 **Don't** copy somebody else's work.

4 **Don't** spend a lot of time designing an elaborate cover. It takes a lot of time and scores no marks.

5 **Don't** put your work in a presentation cover or ring binder. It is best to staple the work together and place it in a polythene pocket.

6 **Don't** forget to check your work is in the correct order before handing it in.

For grades A and A*

EXTENDING THE ORIGINAL PROBLEM

In order to obtain the highest grades at GCSE, you will need to try to extend the problem further. Your will need to introduce several variables, analyse fully alternative approaches and, maybe, explore independently an area of mathematics that is new to you (marks 7 and 8 of the criteria).

In order to do yourself justice and extend your investigation properly you will need to manage your time carefully.

EXPLORING A NEW AREA OF MATHEMATICS

This is actually quite difficult to do. It is also important that the task you choose provides you with opportunities to explore and discover new areas of mathematics. In order to investigate a new area of mathematics which relates to your coursework properly, you will need to have access to A-level mathematics textbooks. Some useful A-level topics include:

● arithmetic and geometric progressions

● differentiation.

Before embarking on the new area of mathematics, make sure you double-check with your teacher.

Tips for carrying out mathematical investigations

Practical work

So far, we have only considered an investigational piece of coursework. It may be the case that you will be asked to carry out a practical piece of work. This type of coursework usually involves doing some research, for example, taking measurements and then analysing the results. All coursework, whether investigational or practical, is marked according to the same criteria. Follow a procedure similar to that for investigational work.

1 Always read the question carefully, make a plan of what you are going to do and what materials you need.

2 Decide how you are going to break the problem down into simpler parts.

3 Collect the data you need from a variety of sources. Make sure you give reasons why you have collected your data and what you want to do with the data.

4 Use calculators and computers in your work whenever you can. If you use a computer, make sure you have explained what programs you have used.

5 Use tables to display your results. Consider various ways of illustrating your results. Always comment on your use of tables, graphs and diagrams.

6 Form conclusions or solutions based on your findings and relate these findings back to the original task or hypothesis. Remember to analyse fully what you have found.

7 Once you have finished the original problem, think of alternative ways in which to extend the problem. Try to develop a new area of mathematics.

Tips for carrying out mathematical investigations

Using examples in practical work

The best way to illustrate the procedures involved in carrying out a practical piece of work is to consider an example. This example goes into considerably less detail than would be expected but it illustrates the key stages of development which you will need to adopt.

Look at this problem.

Max box

Find the maximum volume of a box that is made by folding a piece of card 20 cm × 20 cm square.

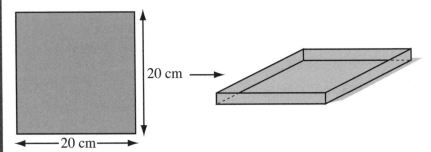

20 cm

20 cm

Although this problem seems fairly straightforward, it would be sensible to try actually making the box by cutting out corners of different sizes from a piece of paper 20 cm × 20 cm square.

Start by making a plan that includes the materials you will need:

● paper

● scissors

● calculator and/or access to a computer.

Tips for carrying out mathematical investigations

Extend your plan by noting how you will tackle the problem, and in what order you will carry it out.

- Start by cutting out pieces of paper 20 cm × 20 cm square.

- Cut squares of side 1 cm, 2 cm, 3 cm, 4 cm, 5 cm … from the corners, fold up the net you have formed and work out the volume of the box each time.

- Put you results in a table.

- Draw graphs of the data.

- Look for any conclusions.

- Check any conclusions.

- Change the size of the original square.

BREAKING THE PROBLEM DOWN INTO SIMPLER PARTS

In order to find the maximum volume of a box formed from a piece of card 20 cm × 20 cm square, it is best to be systematic in your approach. By cutting off squares of side 1 cm, 2 cm, 3 cm, … from the corners and then finding the volume of the box each time, you are breaking the problem down into simpler parts. This systematic approach should help you find conclusions more quickly.

A practical project

Try these steps.

1 Remove squares of size 1 cm × 1 cm.

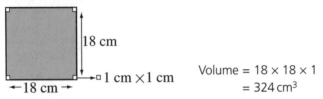

18 cm

18 cm

1 cm × 1 cm

Volume = 18 × 18 × 1
= 324 cm³

2 Remove squares of size 2 cm × 2 cm.

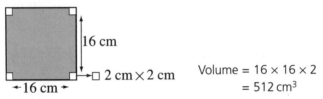

16 cm

16 cm

2 cm × 2 cm

Volume = 16 × 16 × 2
= 512 cm³

3 Remove squares of size 3 cm × 3 cm.

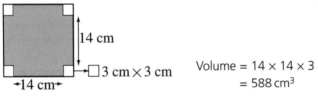

14 cm

14 cm

3 cm × 3 cm

Volume = 14 × 14 × 3
= 588 cm³

4 Remove squares of size 4 cm × 4 cm.

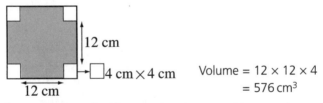

12 cm

12 cm

4 cm × 4 cm

Volume = 12 × 12 × 4
= 576 cm³

So far you have worked in an ordered way, making good use of diagrams to explain how you have worked through the problem.

Continuously drawing the diagrams, however, is very time-consuming, so after monitoring what you have done, stop drawing out the diagrams and cutting out the squares practically.

Tips for carrying out mathematical investigations

Recording and tabulating results

Communicating the results is very important. Placing the **Max box** results in a table shows that the volume increases to a maximum when the square removed is 3 cm × 3 cm (i.e. the height of the box is 3 cm), and then it begins to decrease when the height of the box is 4 cm.

Height of square removed (cm)	Length of the square base (cm)	Volume (cm³)
1	18	324
2	16	512
3	14	588 ⎫ The maximum volume
4	12	576 ⎭ occurs around here.

Based on the observations recorded in the table, it appears that the maximum volume occurs when the size of the square removed has a height between 3 cm and 4 cm.

Since you will need to show several forms of communication for **Strand 2**, it may also be useful to present the tabulated results in a graph. This would help you explain the position of the maximum volume.

Notice that the graph has been **annotated**. This really helps to explain how it supports your results.

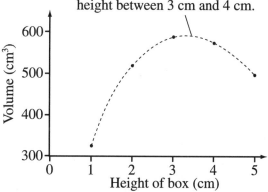

The maximum volume seems to lie between these two points, at a height between 3 cm and 4 cm.

Since the maximum volume has been identified as occurring when the height of the box is between 3 cm and 4 cm, this is now worth investigating. Once again, it is important to be systematic. Increasing the height of the box in steps of 0.1 cm each time would be an ideal way to proceed.

Using ICT: spreadsheets

It would be very time-consuming to continue with the **Max box** problem manually. The use of a **spreadsheet** is ideal for this type of problem: not only will it save you lots of 'number-crunching', but it would also show how you are able to make decisions and show your skills in ICT. However, if you do use a computer, it is important that you explain how and why.

If you work through the problem, increasing the height by 0.1 cm each time, you should find the results that have been tabulated here.

Height of square removed (cm)	Length of the square base (cm)	Volume (cm³) (to 2 decimal places)
3.0	14.0	588.00
3.1	13.8	590.36
3.2	13.6	591.87
3.3	13.4	592.55
3.4	13.2	592.42
3.5	13.0	591.50
3.6	12.8	589.82
3.7	12.6	587.41
3.8	12.4	584.29
3.9	12.2	580.48

The maximum volume occurs around here.

Representing the **Max box** data in a graph would again show that the maximum volume lies between two values, when the height of the square is between 3.3 cm and 3.4 cm. You could then continue the above procedure but this time working with two decimal places, i.e. 3.30, 3.31, ..., 3.39. If you were to continue with this, you should find that the maximum value occurs when the height of the square removed is 3.$\dot{3}$ cm. This gives a maximum volume of approximately 592.59 cm³ (to 2 decimal places).

Although you now know the answer, full credit would only be given if you have shown thorough appreciation of the problem and proof that you have carried out the investigation.

Tips for carrying out mathematical investigations

Continuing the investigation

You have investigated the maximum volume when the side of the square is 20 cm. You could extend the problem to find a general result connecting the height of the square cut out (the height of the box) and the maximum volume.

You could try finding the maximum volume of boxes made from squares of side 10 cm, 12 cm, 15 cm and 24 cm. In fact, any size, up to 100 cm, would be suitable. Try to keep thing simple!

When you have found sizes of cut-out squares and maximum volumes for squares of other sizes, you could analyse your overall results, as **Strand 3** of the marking criteria involve justification and mathematical reasoning. Here are some typical results.

Size of card (s cm)	Size of square cut out (h cm)	Maximum volume (v cm³)
20 × 20	3.$\dot{3}$	592.59...
10 × 10	1.$\dot{6}$	74.07...
12 × 12	2	128
18 × 18	3	432

Analysing the results from the table should help you to make a general rule which connects the height (h) and the side of the square (s), which will always give you the maximum volume (v).

After making the rule, remember to predict and test. *Can you show a further example, to prove that your rule is correct?*

EXTENDING THE PROBLEM

After reaching a successful conclusion with the squares of card, an obvious extension would be to use rectangular pieces. This would introduce several variables. Decide on the size of the rectangular card sensibly, perhaps making it 1 cm longer than it is wide. Be sure to explain how and why you decided on this.

The overall formula which connects the maximum volume, (v), with the size of square removed, (h), and the height of the rectangle, (s), is very difficult to find just from a table of results. However, this practical problem does lend itself very well to an investigation of a new area of mathematics (**Strand 1** mark 8).

It would be very useful if you had access to an A-level textbook that covers differentiation.

A close look at AO4: Handling data

Assessment objective 4 (AO4) is known as *Handling data*. You are required to produce a project, using knowledge, skills and understanding in handling data. This project forms part of your coursework assessment and accounts for 10% of the final mark. Syllabuses for GCSE vary a little in the way that AO4 is assessed. However, all AO4 work will be assessed using the same criteria and the same scale of marking.

The assessment criteria for the *Handling data* project is sub-divided into three areas:

- **Strand 1 – Specify the problem and plan**: this strand is about how you plan your project, what type of data you collect and how you will analyse it.

- **Strand 2 – Collect, process and represent data**: this strand is about the different ways in which you choose your data and the variety of statistical methods you use with your data.

- **Strand 3 – Interpret and discuss results**: this strand is about the conclusions that you draw from your data and how they relate to the statistical diagrams and measures that you have used.

Mark descriptions comprising a number of statements are provided for each area of the project. Each of these strands is assessed and marks are awarded out of a maximum of 8. Unlike AO1 the descriptions are given in mark bands, 1–2, 3–4, 5–6, … within each strand.

The table on pages 76 and 77 shows the basics involved in these criteria.

MARKING

When marking your project, your teacher or the examination board needs to find the band which best fits your work in each of the three strands. Once they have found the band, they allocate a mark. The lower of the marks is given if you have just satisfied the criteria.

The data-handling cycle

The **data-handling cycle** is very useful when doing the *handling data* project. You should already have seen and used the data-handling cycle while working on mini-projects during Key Stage 3.

The cycle has four parts to it. You will find it useful to refer to the cycle when you are planning your project.

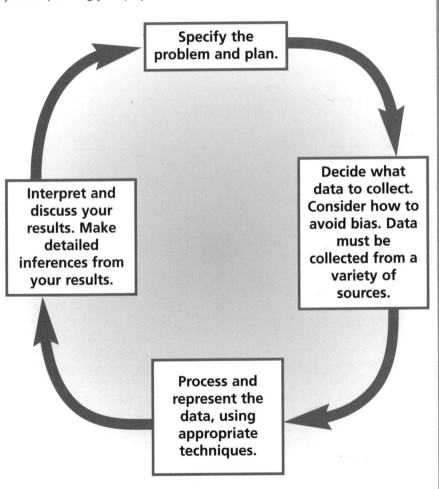

Specify the problem and plan.

Decide what data to collect. Consider how to avoid bias. Data must be collected from a variety of sources.

Process and represent the data, using appropriate techniques.

Interpret and discuss your results. Make detailed inferences from your results.

Handling data
Assessment criteria

Marks	Strand 1: Specify the problem and plan	Strand 2: Collect, process and represent data
1–2	I can choose a simple problem and explain what I am going to do and how I am going to do it.	I can collect some data and I can make some simple calculations. I can draw some diagrams and I will present my results in a clear and organised way. I try to relate the results to my plan.
3–4	I can write a plan explaining what statistical techniques I am going to use. I write out reasonably clear aims. My sample is an adequate size.	I can collect relevant data and make relevant calculations of Grade F standard. I can use statistical terms in my write-up and I am able to present information in a variety of forms, using suitable scales and titles. I will refer to my results and diagrams.
5–6	I can consider a more complex problem. My plan is well written, using statistical terms, and I have considered possible problems. I can explain my choice of sample. My report is well set out and it all makes sense.	I can collect relevant and reliable data and I can make relevant calculations of grade C standard. I have not missed out any obvious calculations and all my calculations were used. I can explain my results throughout the project, using statistical language. I am able to explain why I have chosen particular methods of presentation.
7–8	The problem that I am considering is very complex and requires me to specify carefully how I am going to solve it. I can write my plan in statistical language and I can give reasons for each decision made. I can plan for possible problems which may arise. I am able to explain how I chose my sample and how I tried to avoid bias. The written report is well structured and the conclusions relate to the aims.	I collected relevant and reliable data and I can make relevant calculations of grade A standard. No obvious calculations will be missed out. All my calculations will be used and of suitable accuracy. All presentation has purpose, which I will explain. I will deal with problems such as nil response or missing data.

Strand 3: Interpret and discuss results

I can comment on the patterns in the data and summarise the results.

I can comment on the data and I will note any exceptions.
I can give correct interpretations of my graphs and calculations.
I am able to relate my summary to my initial problem and I will try to explain how well I did.

I can comment on the data and give reasons for exceptions.
I can interpret my graphs and calculations correctly.
I can explain the significance of my results, taking into account my sample size.
I am able to relate my summary to my initial problem.
I am able to evaluate the effectiveness of my overall strategy and discuss any limitations.

I can comment on the data and give good reasons for exceptions.
I can make correct and detailed interpretations of my graphs and calculations.
I am able to talk about the likelihood and significance of my results, taking into account bias and sample size.
I have related my summary to my initial problem and explained how well I did with suggestions for improvement.
I can comment constructively on the practical consequences of my work.

Choosing a project

If your teacher sets a project for you, move on to the next section.

If you have to choose a project for yourself, make sure that it is one that interests you. There are many different ideas to choose from. Some suggestions are given below.

HOBBIES OR INTERESTS
- Types of books people read
- Types of television programme
- Newspaper comparisons
- People's memories
- Opinion surveys about, for example, sports centres, local parks

WHERE YOU STUDY (SCHOOL)
- Examination results, for example, GCSE, KS3
- Comparisons of statistics such as heights or weights of students in different years
- Reaction times, comparing those of boys and girls
- Absence rates
- Lateness
- Hours spent on homework

LOCAL OR NATIONAL DATA
- Weather
- Traffic
- Census data

Tips for carrying out a handling data project

Formulating a hypothesis

When you have decided on the project, write down some ideas that you could test. Choose one and write it as a statement. This is your **hypothesis**.

Choose a hypothesis that allows you to use a variety of statistical techniques. It is important to give reasons as to why you have chosen this hypothesis.

In order to achieve higher marks on your project you will need to make several hypotheses and investigate each one.

The purpose of making a hypothesis is to enable you to investigate whether you are right or wrong. Some examples of hypotheses are given below.

STUDENTS
You may decide to investigate the relationship between heights and weights of some students. Possible hypotheses include:
- The taller the student the more they weigh.
- Height and weight are positively correlated.
- The correlation between height and weight decreases with age.
- Students in years 10 and 11 have better correlation between height and weight than students in years 7, 8 and 9 do.

NEWSPAPER COMPARISONS
You might compare word lengths and sentence lengths between tabloid and broadsheet newspapers. Possible hypotheses include:
- Broadsheet newspapers have longer words than tabloid newspapers.
- Broadsheet newspapers have longer sentences than tabloid newspapers.
- Broadsheet newspapers contain more text than pictures.

CARS
You could investigate factors that influence the price of a second-hand car. Possible hypotheses include:
- The older the car, the less expensive it will be.
- The higher the mileage, the lower the price of a second-hand car.
- The higher the price of the car when new, the higher the second-hand price.

Making a plan

It is important that you give a well-structured plan of how you intend to carry out the task. You have **chosen your project** and **written your hypothesis**. Now you need to **plan ahead** and think about what **data to collect**, how to collect your data and what to do with the data you have collected.

These checklists suggest some questions for you to think about when you are at this stage in your planning. You will need to refer back to some of the questions as you work through your project.

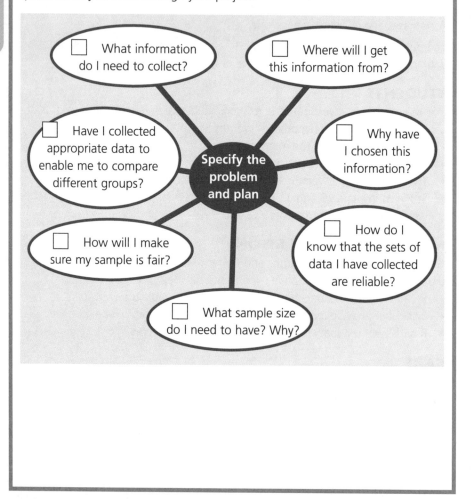

Tips for carrying out a handling data project

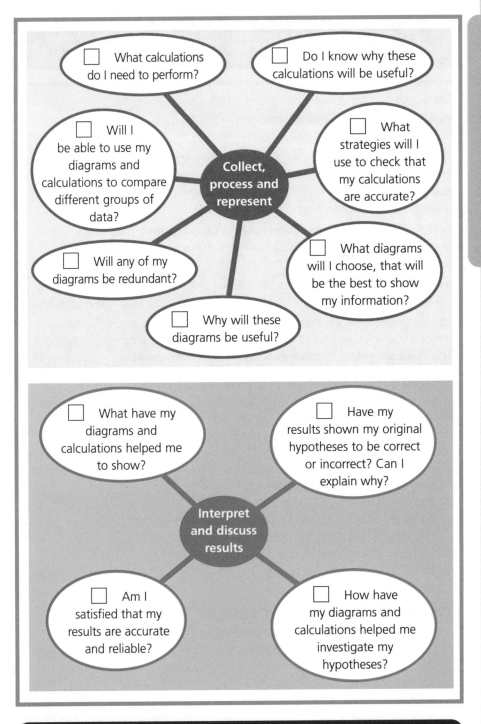

What calculations do I need to perform?

Do I know why these calculations will be useful?

Will I be able to use my diagrams and calculations to compare different groups of data?

Collect, process and represent

What strategies will I use to check that my calculations are accurate?

Will any of my diagrams be redundant?

What diagrams will I choose, that will be the best to show my information?

Why will these diagrams be useful?

What have my diagrams and calculations helped me to show?

Have my results shown my original hypotheses to be correct or incorrect? Can I explain why?

Interpret and discuss results

Am I satisfied that my results are accurate and reliable?

How have my diagrams and calculations helped me investigate my hypotheses?

Tips for carrying out a handling data project

Collecting the data

There are two types of data.

Qualitative data items include things that are descriptive such as eye colour, make of car.

Quantitative data are numerical data and they may be either **discrete** or **continuous**. Discrete data involve counting, for example, marks in an examination or number of children in a family, while continuous data involve measuring, for example height, weight or length.

Data that you collect yourself are known as **primary data**. Data that you are given or that you download from the internet or a database are known as **secondary data**.

It is important that you choose a suitable sample and sample size. You need to collect at least 30 sets of data and each set must have at least two aspects to it.

You also need to decide how to collect the data and what type of sampling method you will use. There are lots of different sampling methods, including **stratified sampling**, **quota sampling** and **random sampling**.

USING THE RANDOM NUMBER BUTTON

There are different ways of taking a random sample. For example, if you have 400 boys and you wish to have a sample size of 30, you could write each student's name on a piece of card and choose 30 without looking. This would take a long time. An alternative would be to give each boy a number from 1 to 400 and use the random number button on your calculator to choose a sample.

- Generate a random number between 0 and 1.
  Press SHIFT RAN.
- Multiply the number by 400 to obtain a number from 1 to 400.
- Round your answer to the nearest whole number.
- Carry out the process 30 times to select a random sample of 30 boys.
- If the calculator selects a number you have already chosen, ignore it and try again.

CARRYING OUT A PILOT SURVEY

It is sometimes useful to carry out a **pilot survey** before embarking on the project. This will help you check that the data you have collected will help you prove or disprove your hypotheses. If not, then you will need to alter your survey appropriately.

Tips for carrying out a handling data project

Recording the data

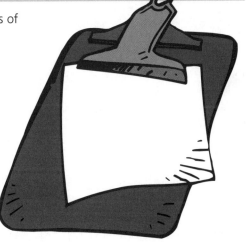

Once you have collected your data you will need to record your data in a table. Use a table that is clear and easy to read.

EXAMPLE 1

Make of car	Price when new (£)	Age (years)	Price when second-hand (£)
Mazda	10 590	7	3592
Vauxhall	7450	6	3495

EXAMPLE 2

Gender	Height (m)	Weight (kg)
Male	1.77	59
Female	1.66	45
Female	1.60	56

Obviously, you can use other types of table such as **tally charts** and **grouped frequency tables**.

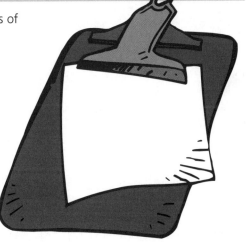

Tips for carrying out a handling data project

Presenting the data

Once you have collected and recorded the data you need to present your results, using suitable graphs and charts.

It is important that you choose the most appropriate methods to present your data. For example, to look for relationships, use scatter diagrams. To compare distributions, use box plots or cumulative frequency diagrams.

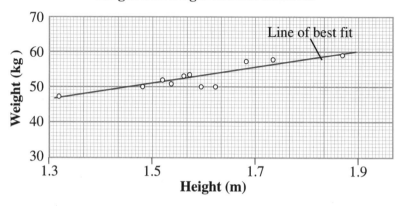

Height and weight of some students

Height (m)	Weight (kg)
1.63	50
1.57	55
1.47	50
1.6	50
1.54	52
1.86	58
1.32	47
1.68	57
1.58	55
1.52	54
1.74	57

This scatter diagram can be used to show the correlation between the heights and weights of some students.

The table on the opposite page outlines some of the types of mathematics you could use in order to achieve marks in Strand 2. It is important to realise, though, that drawing a scatter diagram does not automatically mean that you will get a mark of 4. The content of the diagram must be appropriate.

Analyse and interpret

Mark	Processing and representing
1	Draw bar charts and pictograms. Calculate the mode and range of data.
2	Draw line graphs. Calculate the mode of the grouped data.
3	Draw simple pie charts and stem-and-leaf diagrams. Calculate the mean and median of discrete data.
4	Draw frequency diagrams for continuous data. Draw scatter diagrams and explain the correlation. Draw more difficult pie charts.
5	Draw frequency polygons and the line of best fit on a scatter diagram. Calculate the mean, median and modal class of grouped data.
6	Draw cumulative frequency graphs and box plots. Use the cumulative frequency graphs to find the median and interquartile range. Draw time-series graphs and calculate the moving averages.
7	Draw and compare histograms.
8	Understand and use correlation coefficients and standard deviation.

Tips for carrying out a handling data project

Analyse, interpret and draw conclusions

ANALYSE AND INTERPRET

To enable you to analyse your data, it is important that you have selected and carried out appropriate calculations. These may involve calculations involving items such as **central tendency**, **dispersion**, **correlation** or **time series**.

It is now essential that you analyse your data, interpret the outcomes and discuss the results and findings. When summarising the results, you will need to refer back to all of the graphs and comments. If you have several hypotheses it would be useful to interpret and discuss your findings separately for each hypothesis. When writing your report you may wish to follow this outline.

- State the relationships in what you have found.

- Write detailed explanations about your interpretation of the diagrams, charts and calculations.

- Make comparisons of and between your data.

- Decide whether your results support or reject your hypothesis and explain why.

FINAL CONCLUSIONS

To complete your project you must make some final conclusions.

- Make a list of points that summarise your findings.

- Explain how successful you were in achieving your aims.

- Discuss any anomalies.
  Where there any problems or limitations with the project?

- Relate your findings to the real world.

At this stage you could also include some ideas for further investigation that would extend and enhance your project.

For grades A and A*

In order to achieve Grades A and A* you need to fulfil the criteria in all three strands for marks 7 and 8 (see page 85).

It is essential that you:

- choose a problem which is complex enough and which allows you to use a range of higher-level statistical techniques
- specify carefully and write out a clear plan – you must be able to explain all the decisions you have made and plan for any possible problems
- explain how you have collected the sample and how you intend to avoid bias
- can make correct and detailed interpretations of calculations and graphs
- can talk about the significance of the results, taking into account bias and sample size.

The statistical measures below are beyond the *Handling data* section of the GCSE specification. However, you may find them useful when analysing data at the higher level. If you do use these measures you must show that you understand what they do and that you can interpret the results correctly.

STATISTICAL MEASURE

Standard deviation used to measure the spread of data.

Product moment correlation coefficient a measure of how good the relationship between two sets of data is. You would use this with a scatter diagram.

Skewness a measure of the spread of the data on either side of the mean.

Rank correlation a way of comparing two lists of data.

Using ICT

When carrying out your *handling data* project, sorting the data can take a long time, and this may be before you have even started drawing 50 sets of data on a scatter diagram.

The task can be made a lot easier if you use a spreadsheet to process and analyse your data. Spreadsheets will do lots of statistical analysis in a fraction of the time that it takes a student.

You will not lose marks for the use of ICT; in fact, the correct use of ICT is encouraged by the examination boards.

Listed below are some of the analysis tasks that a spreadsheet can do.

Sorting   for putting data in order when you want to work out medians and quartiles

Representing data   for drawing statistical diagrams such as bar charts, pie charts, scatter diagrams

Functions   for working out statistical information such as:
- correlation coefficient
- upper and lower quartiles
- median
- mean
- the equation of a line of best fit
- the largest and smallest value of the data
- standard deviation
- skewness

Other statistical diagrams such as **frequency polygons**, **box plots**, **histograms** and **cumulative frequency diagrams** will need either to be drawn by hand or to be done on specifically-designed statistical packages.

Using ICT is excellent for saving time on some of the 'number crunching'. Remember, though, that you must fully explain what you have done, and why. A project full of diagrams and calculations with no reasoning or explanation would score few marks.

Tips for carrying out a handling data project

Writing up coursework

By now, you have worked through the procedures that are needed in order to carry out an investigative piece of coursework and the *Handling data* project. Writing up your project is extremely important since it is the only evidence available for comparing your work with that of other students. Writing the best possible report is obviously very important. Here are some points to remember.

- The report you write can be handwritten or word-processed. Most are handwritten. If you handwrite your work, write as clearly as possible. Your report is going to be read by other people who moderate the work. They must be able to read your writing, otherwise some of the important points that you are trying to make will be missed. It is often helpful if you use a different coloured pen or a highlighter to pick out the points you think are important, such as the formulae or the number patterns. This will draw attention to what you think is significant.

- If your report in word-processed, you must do it yourself. Choose a font that is easy to read. There are fonts that are more difficult to read than handwriting, and some may not include all the characters that you need for mathematics. Do be careful to include full explanations, decisions made and good justifications. Surprising as it may seem, students who use word-processing are often tempted to write these in a very brief form.

- Don't be tempted to draw an elaborate front cover, as it will score you no extra marks. Try to write up your report as you proceed, otherwise your work will be rushed at the end and, quite possibly, not justify your findings as fully as you want.

- Before handing in your project, always have a final check that the work is in the correct order (numbering pages would help), all diagrams are labelled and your name is on it. It is helpful to place your work in a polythene pocket.

- Your coursework must be presented in your own words. If you copy from a book or from somebody else then your work would not be admissible. If you are given the *Max box* investigation, you **must not** copy from this book – use it only for reference!

Your final mark

USING AND APPLYING TASK

Your teacher is required to award you a mark from 1 to 8 for each of the three strands of the AO1 criteria. Each of these strands should represent your best performance within a strand across the tasks submitted. (You are allowed to submit two tasks.) Your marks should be totalled to give a final mark for the task out of a possible 24.

HANDLING DATA PROJECT

Your teacher is required to award you a mark from 1 to 8 for each of the three areas of the AO4 criteria. Your marks should be totalled to give a mark for the project out of a possible 24.

The marks for the task and the project should be added together to give a final overall total mark for coursework out of a possible 48.

Task 1		Task 2 (Optional)		Project	
Strand	Mark	Strand	Mark	Area	Mark
1		1		1	
2		2		2	
3		3		3	

To calculate your final mark, write down your marks for each of the three strands for the two coursework tasks you have completed. (You may have only completed one task.) Take the highest mark from each strand and add these up to give you a mark out of a possible 24.

Now write down your marks for each of the three areas of the *handling data* project. Add up your marks to give a total out of a possible 24.

Add the mark for the task and the project together to give a mark out of a possible 48. This is your final coursework mark.

Glossary

Algebra The use of letters to represent numbers.

Angle A measurement of turn.

Approximation A rough answer. Approximations are made by rounding off information.

Arc Part of the circumference of a circle.

Area The amount of flat space covered by a 2-D shape.

Average speed The total distance travelled divided by the total time taken.

Axis The horizontal or vertical line on a graph from which coordinates are measured.

Bar graph A diagram made up of a set of bars, of equal width. The lengths are proportional to a set of frequencies.

Bearing An angle measured from north in a clockwise direction. Bearings have three figures.

Bias A tendency in an event to favour one outcome rather than others.

Bisect Cut exactly in half.

Bisector A line which divides a line, angle or area exactly in half.

Brackets Characters or symbols used to show terms that should be treated together.

Capacity A measure of the amount of space inside a 3-D object.

Circumference The distance around the outside edge of a circle.

Class interval A grouping of statistical data.

Coefficient The number in front of a letter, in an algebraic expression.

Common factor A factor which is the same for two or more numbers.

Complementary angles Angles that add up to 90°.

Congruent shapes Shapes that are the same size and the same shape.

Continuous data Data obtained by measurement.

Cumulative frequency A running total of all the frequencies in a data set.

Data The collective name for pieces of information, often obtained from an experiment or survey. *Data* is a plural noun.

Degree A unit for measuring angles.

Denominator The number on the bottom of a common fraction.

Density The mass per unit volume of a substance.

Diagonal A line joining any two non-adjacent corners of a shape.

Diameter A straight line that passes through the centre of a circle and cuts the circumference in two places.

Discrete data Data that can be counted.

Edge The line where two faces meet in a 3-D shape.

Elevation The view of a 3-D shape from its front or side.

Equation A method of writing two or more things that are equal.

Equidistant A point that is the same distance from 2 or more points or lines.

Evaluate Work out the value of an expression.

Exterior angles The angles formed on the outside of a polygon when the sides are extended.

Face A flat surface of a 3-D shape.

Factor A number that divides exactly into another number.

Factorise Separate an expression into its factors.

Formula A mathematical expression that is used to calculate an unknown when values are given.

Frequency The number of times that an outcome or event occurs.

Frequency density The height of a bar on a histogram.

Frequency polygons The shape formed by joining the midpoints of class intervals in a histogram for grouped or continuous data.

Gradient The slope of a line in relation to the positive direction of the x-axis.

Histogram Similar to a bar chart, except it has no gaps and is used for showing continuous data.

Horizontal Straight across, parallel to the earth's surface.

Hypotenuse The longest side of a right-angled triangle.

Hypothesis A statement that can be tested to see if it is true or false.

Independent events Events that have no effect on each other.

Index The power to which a quantity is raised.

Inequality A statement that two or more things are not equal.

Integer A positive or negative whole number or zero.

Intersection The point where two lines meet.

Intercept The point at which a line cuts an axis.

Interior angles The angles inside the corners of a polygon.

Irrational numbers A number which cannot be written as a fraction.

Isosceles triangle A triangle where two sides and two angles are equal.

Linear A line, in one dimension. A linear graph is a straight line. The graph of a linear equation, such as $y = 2x + 3$, is a straight line.

Line graph A graph, formed by joining plotted points.

Locus The set of all possible positions of a point that moves subject to a given condition or rule.

Lower bound The smallest number that may be included in the set.

Lowest terms A fraction is in its lowest terms when it cannot be cancelled any further.

Magnitude The length of a vector.

Mapping A relationship between one group of numbers and another group of numbers.

Mean A type of average, found by dividing the sum of a set of values by the number of values in the set.

Median A type of average, found as the middle value when a set of numbers is put in order of size.

Mode A type of average. The most frequently-occurring number or value in a set.

Multiples The numbers in the multiplication tables. For example multiples of 5 are 5,10,15, 20, ... since 5 will divide exactly into these numbers.

Multiplier Scale factor.

Mutually exclusive events Events that cannot happen at the same time.

Net A flat shape that can be folded to form a 3-D solid shape.

Numerator The top part of a common fraction.

Outcomes The possible results of a statistical experiment or other activity involving uncertainty.

Parallel lines Lines that never meet; they are always the same distance apart.

Perimeter The distance around the outside edge of a shape.

Percentage A fraction with a denominator of 100.

Perpendicular Meeting at 90°.

Plan The view of a 3-D shape, seen from above.

Polygon A plane figure with three or more straight sides. A regular polygon has all sides equal and all angles equal.

Prime number A number that has exactly two factors: 1 and itself.

Product The result of multiplying two or more numbers together.

Quadratic equation An equation that contains a squared term.

Quadrilateral A polygon with four sides.

Quartile One of three values that divides a frequency distribution into 4 levels. The lower quartile is at $\frac{1}{4}$ level, the median is halfway and the upper quartile is at $\frac{3}{4}$ level.

Questionnaire A sheet with questions, used to collect data.

Radius The distance from the centre of a circle to its circumference.

Range The difference between the highest and lowest values in a set of data.

Ratio A comparison between two quantities that are measured in the same units.

Reciprocal The reciprocal of a number is the result of dividing 1 by it. The reciprocal of $\frac{a}{x}$ is $\frac{x}{a}$.

Relative frequency An estimate of probability.

Scale factor The multiplier when a shape is enlarged or reduced in size.

Sequence A set of numbers that follow a pattern or rule.

Similar Two or more figures are similar when they are exactly the same shape but not necessarily the same size; one is an enlargement of the other.

Simplify Make a numerical or algebraic expression easier to understand.

Simultaneous equations Two or more equations that connect the same two, or more, unknowns and have a common solution.

Standard index form A way of expressing very large or very small numbers in the form $A \times 10^n$ where $1 \leqslant A < 10$ and n is an integer.

Substitution Replacing a letter with its numerical value.

Tangent The tangent of an acute angle in a right-angled triangle, is the ratio of the side opposite the given angle to the adjacent side. A tangent to a circle or curve is a straight line touching it at one point.

Term A part of an expression.

Triangle A polygon with three sides.

Trigonometry The study of triangles and the relationship between their sides and angles. Trigonometric ratios, i.e. sine, cosine and tangent are used.

Upper bound The number which no number in the set exceeds.

Variable A quantity that can take on a range of values.

Vector A quantity that has both size and direction.

Vertical Straight up and down, at 90° to the horizontal.

***x*-axis** The horizontal (along) axis.

***y*-axis** The vertical (up) axis.

Glossary

Every effort has been made to contact the holders of copyright material, but if any have been inadvertently overlooked the publishers will be pleased to make the necessary arrangements at the first opportunity.

Published by Letts Educational
The Chiswick Centre
414 Chiswick High Road
London W4 5TF
tel: 020 89963333
fax: 020 87428390
e-mail: mail@lettsed.co.uk
website: www.letts-education.com

Letts Educational Limited is a division of Granada Learning Limited, part of Granada Plc.

Text © Fiona Mapp.
Design and illustration © Letts Educational Ltd 2005.

First published 2005

ISBN 184315 4706

The author asserts her moral right to be identified as the author of this work.

British Library Cataloguing in Publication Data

A catalogue record for this book is available from the British Library.

Cover design by Big Top.

Commissioned by Cassandra Birmingham

Project management for Letts by Julia Swales

Edited by Joan Miller

Design and project management by Ken Vail Graphic Design, Cambridge

Printed and bound In Italy.